Chassis & Suspension Engineering

Chassis & Suspension Engineering

Road and Track
Theory and Practice

GEOFFREY HOWARD

OSPREY

First published in 1987 by Osprey Publishing Limited
59 Grosvenor Street, London W1X 9DA

Reprinted Spring 1989

Sole distributors for the USA

Motorbooks International
Publishers & Wholesalers Inc
Osceola, Wisconsin 54020, USA

British Library Cataloguing in Publication Data

Howard, Geoffrey
Chassis & suspension engineering.
1. Automobiles—Chassis
I. Title
629.2'4 TL255
ISBN 0-85045-775-0

Editor Tony Thacker
Design Norman Brownsword

Filmset by Tameside Filmsetting Limited,
Ashton-under-Lyne, Lancashire
Printed by Butler & Tanner Limited,
Frome, Somerset, Great Britain

Contents

Acknowledgements 6
Preface 7
1 Early chassis and suspension design 9
2 Introduction to the science 26
3 Spring and damper theory 34
4 Modern suspension systems 44
5 Drive layout and its effects 61
6 Computer aids and design techniques 82
7 Tyres and their influence 93
8 Racing-car suspension systems 105
9 Active and adaptive ride 127
10 Test and appraisal methods 144
11 Rear-wheel steering systems 153
12 Braking systems and anti-lock 164
13 Future trends 175
Index 188

Acknowledgements

I am deeply indebted to my one-time colleague on *Autocar*, Ted Eves, whose intimate knowledge of Formula 1 racing-car design proved a great help with the chapter on Grand Prix chassis design, to the suspension designers who worked with me daily during my years with the Ford Motor Company and to the countless development engineers throughout the auto industry for their co-operation, assistance and enthusiasm for the subject.

I am also grateful to the following for the provision of illustrations on the pages indicated:

Adam Opel AG (pages 16, 35, 53, 54, 84, 142, 143, 156 and 180), Alfred Teves Gmbh (pages 165, 166, 168, 171 and 186), Audi AG (pages 24, 71, 72, 145 and 146), Austin Rover Group (pages 39, 68 and 129), Automotive Products (page 173), Bilstein (pages 59 and 184), Continental Tyres (pages 102 and 103), Daimler-Benz Museum Archives (pages 9, 20, 22 and 52), Edward Eves Archives (pages 106, 108, 110, 113, 114, 117, 119, 122 and 124), Firestone Tyres (page 94), Ford Motor Company (pages 18, 27, 40, 41, 42, 45, 49, 62, 83, 85, 86, 87, 89, 120, 148, 167, 178, 185 and 186), Honda Motor Company (pages 153, 159, 160, 161 and 162), Ital Design (page 158), Jaguar Cars (pages 48 and 51), JJD (pages 98 and 100), Lancia SpA (pages 72, 78 and 79), Lotus Cars (pages 136 and 137), Lucas Girling (page 172), Mazda Ltd (pages 47 and 56), Monroe (page 59), National Motor Museum (page 12), Nissan Motors (pages 141, 157 and 176), Peugeot-Talbot (page 57), Pininfarina Industrie SpA (page 15), Porsche Cars (pages 155 and 156), Quadrant Autocar Archives (page 10), Regie Renault (page 64), Saginaw Steering (page 183), Steven Gooderham (page 14), Vauxhall Motors Ltd (page 47), Volkswagen AG (pages 46, 57, 76 and 77), Volvo AB (pages 132, 140, 146, 147 and 149), ZF Gmbh (page 182).

Preface

Few areas of vehicle design are shrouded in as much mystery as the dynamics of motion, and all that the term entails. Ride, handling, steering and braking are all complex and crucial characteristics in the subjective man/machine relationship of any driver-controlled vehicle, be it something as simple as a bicycle or a full-blown, 850 hp, Formula 1 racing car.

This book is intended to explain the basics of chassis and suspension systems and to probe some of the engineering advances being made for use on road and track. To do that it tracks the development of vehicle design, describes what happens to the wheels on bumps and in turns, investigates some of the systems and computer aids being used in design and development, and looks at what the future holds.

The term 'chassis' is, by convention, applied to both the original structure which supported the running gear of early vehicles and the subsequent range of components associated with it—the axles, springing and damping media, locating links, steering and braking systems. 'Suspension' is confined more precisely to the systems between the wheels and the vehicle's occupants—everything from the insulating bushes to the seat upholstery.

Ask an average driver to explain his car's suspension system, or even describe precisely its ride or handling, and you will probably draw a blank response or, at best, a fairly muddled reply punctuated with technical terms he doesn't understand. Most people are unfamiliar with the engineering detail responsible for what they experience behind the wheel. But they usually know exactly when they like the 'feel' of a car and equally when they do not. Although the senses which appreciate being in control of a car that gives a reassuring feeling of confidence, well insulated from harsh disturbances, are extremely subjective and hard to communicate, they are forming an increasingly important consideration in new car design and are a key competitive issue in the market as overall standards improve each year.

To understand the complex dynamics of vehicle movement, some basic facts must be learnt, some principles grasped and a lot of information taken for granted. Therefore, this book is aimed at leading the reader along a gentle learning curve that will remove much of the mystery from the subject and improve the ability to assess and appreciate all types of car on the road.

In many ways, this book is dedicated to a band of engineers with whom I shared four long days of driving in the winter of 1980. We covered more than 2500 miles (4000 km) over every kind of road surface Europe had to offer in a convoy of Ford Escorts, each with a different suspension setting, appraising, comparing, making judgements and sitting well into the night at debriefing sessions. The experience taught me how to assess suspension systems by the seat of my pants, by the jerks on my neck muscles and by a series of road manoeuvres unknown to the layman.

Like my previous book *Automobile Aerodynamics*, this work is both a text book and an enthusiast's guide. It cannot tell the whole story of chassis and suspension design, but it should fill in most of the blanks for those who take the trouble to read it.

Geoffrey Howard
Cambridge
March 1987

1 Early chassis and suspension design

From the birth of motoring, sometime around 1885–86, to the end of the nineteenth century and even into the next, there was no well-defined direction for car design to take. At the time, communications were sparse and unreliable, even across continents, and many automotive pioneers worked in secrecy and isolation, developing their own ideas well insulated from outside influences. Consequently, there was a wide variety of methods and styles used and often some irrational mixtures of materials and design.

Having been derived from coach-building traditions, chassis design followed the general layout of wooden carts and carriages. The main frame was usually an assembly of beams, laid out flat, comprising two longitudinal side members braced

Gottlieb Daimler's first motor carriage, built in 1886, used rigid cartwheels and beam axles. The primitive open bodywork was mounted on four sets of double semi-elliptic leaf springs installed back to back

In 1892 Daimler built this ornate motor carriage for the Sultan of Morocco with unsprung bicycle front forks and tiller steering. Seating was still well above the tops of the large rear cartwheels

by riveted cross-members at strategic points along their length. Its dimensions were substantial enough for it to be self-supporting and as stiff as a two-dimensional structure can be, and it was totally independent from the bodywork.

Even before 1900, the excitement and potential of the car was bringing competition motoring to the fore as a means of promoting sales against some firmly-entrenched ideas against internal-combustion engines. In the quest for speed and the spoils of victory, manufacturers developed cars with more power, cars for sport and cars with which to win races.

Front axles were dropped in the centre and swept up at the ends to lower the chassis (and the centre of gravity), while rear springs (universally of the leaf type) were flattened or even reversed to achieve the same effect at the rear. One or two brave innovators, notably Stabilia in France and American Underslung in the USA, even tried mounting the chassis under the axles to reduce the height of the heaviest component.

At first, panelling was mostly in aluminium because the high ductility of this material allowed easy hand-forming. But the invention of the first sheet-metal press tool by Artz in the late 1920s opened the door to steel, with its lower cost and greater strength. Unlike the massive upper and lower dies in today's press tools, the original Artz press was a hydraulic stretcher which pushed a wooden former up from under the panel while it was gripped by clamps at its edges. Similar tools are still used at Aston Martin, but the formers are made from modern resin materials.

The use of steel for body panelling, and subsequently for the body frame itself, was introduced by Daimler with the first Mercedes of 1901. By 1903, there were individual examples of metal-covered bodies in the USA, and in 1905 Pierce-Arrow were using thin-gauge, aluminium castings for some body panels to achieve the complex curvatures required by their designs. Generally, however, the all-steel body was developed by the Budd Manufacturing Company of Philadelphia around 1915, although early steam cars, like the American Victor of 1900, used steel as a fireproof material in the locomotive tradition.

The first large company to adopt the Budd principle, which was based on their experience with railroad cars, was Dodge in the early 1920s. Over the next few years, it became the standard construction method as the industry geared up for ever-increasing volumes of output.

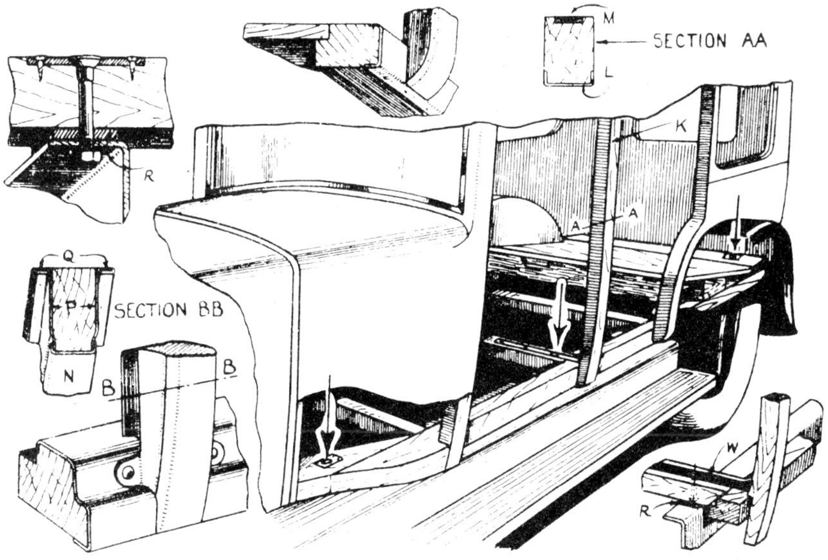

RIGHT *Early coachbuilt bodies were based on a timber frame, reinforced by steel plates and brackets. Metal panels were wrapped round the framework before the trim was added*

BATTALION CHIEF
DIST. No. 5.

The first cars to adopt the Budd principles of the all-steel body without a timber frame were the Dodge tourers of the early 1920s

In 1911, however, the French coachbuilders Carrosserie Henri Labourdette had completed a design exercise which broke new ground in body construction. Under directions from René de Knyff, a successful young racing driver who later became manager at Panhard-Levassor, they built a lightweight body which fell somewhere between a canoe and an aeroplane in appearance. It set a style which was still popular nearly 50 years later.

The brief was to save a considerable proportion of the weight of a standard body by adopting aircraft construction techniques, and the result was a revolutionary design in the form of a tub-shaped shell, without doors, which flowed like a fuselage around the seats and into a long nose and swept-up mudguards. The frame was covered with polished, laminated mahogany strips, copper-riveted in place. It was known as the Skiff, after the elegant, lightweight rowing boat with similar lines and construction. Its appearance caused a sensation and led to many similar styles, including the famous 'boat tail' so popular on sports cars in the 1920s.

The problems of bodywork which flexed on the far-from-rigid chassis of the time led to the development of the Weymann body in France during the early 1920s. It deviated from conventional practice by having a light wooden frame joined by metal strips which allowed the body to flex. The outside was covered with flexible leathercloth and the inside trimmed with cloth, with thick padding between the two layers. It worked on the principle that flexible pin joints can only transmit tension and compression loads, whereas rigid joints (used up to that time) introduce bending moments into the members, which are then subject to fatigue failure. Therefore, the Weymann body allowed much thinner frame sections to be used, while the lack of metal panels also contributed to its extremely light weight.

A fabric-covered style for bodies resulted in America and Europe, although it was not always to the genuine Weymann system, which was protected by patents. At the same time, in 1922, an Italian engineer and one-time racing driver, Vincenzo Lancia, designed a car that was technically years ahead of its contemporaries. This was the amazing Lambda V4.

Apart from advancing suspension design significantly with an independent front system that really proved effective, Lancia's Lambda used deep, pressed-steel side members for its chassis which extended well up towards the waistline to form a frame for the lower half of the body. It was braced across the scuttle and behind the front seats, and provided the kind of rigid frame required for the front suspension to work effectively, thus anticipating the modern, integral steel body by some 15 years.

A few years later, in 1925, André Citroën secured a French licence to build all-steel bodies to the American Budd patents used by Dodge and put a small, very

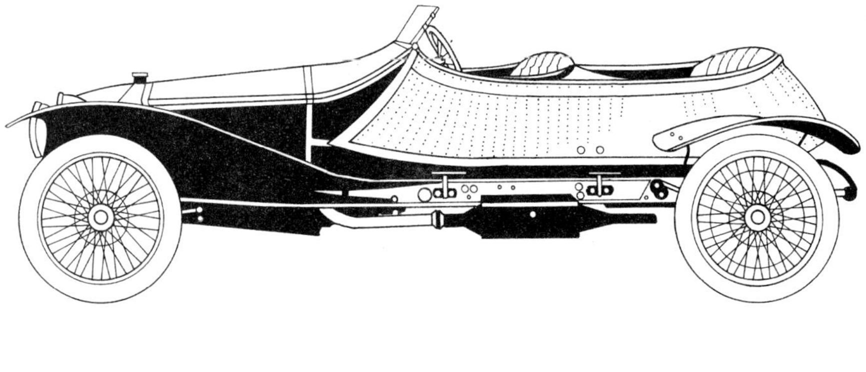

LEFT *Radical innovation in design and construction came in 1911 with this unique 'skiff' roadster constructed by Henri Labourdette in Paris. It resembled a canoe with its thin wooden planks steamed to shape and fastened to the frame by copper rivets*

Equally radical and technically years ahead of its time was the 1922 Lancia Lambda. It used deep pressed-steel side members as the lower half of the open bodywork to provide an exceptionally stiff structure to support its advanced independent front suspension

square saloon into production on the 11 CV chassis. They were produced at the rate of more than 100 per day and sold at a highly competitive price. In the same year, William Morris also came to an agreement with the Budd company and formed the Pressed Steel Company in Oxford.

The significance of an all-steel body, apart from being lighter and cheaper on materials, was that it allowed bulk production of parts and the much faster assembly required by the growing demand for cars.

With the increase in production rates came the need to communicate the mass of technical data required in the design of the body, not the least of which was the actual shape of each panel. Unlike the mechanical systems of the car, the body is extremely complex in shape. The traditional craftsmen, previously responsible only for hand-forming a high-quality body, were now required to become stress engineers and to learn new techniques for the complicated business of reproducing curvature not by eye but from a succession of surface lines on a flat, two-dimensional drawing.

Like timber-framed coachwork, all-steel bodies were detachable at first, following the old manufacturing traditions and allowing easy access during assembly of the chassis and mechanical components as they rolled down the production line. Then it became cheaper and easier to rivet the body to the chassis,

The first integral steel body which dispensed with the traditional chassis was the 1935 Opel Olympia. Similar weight-saving concepts did not become widespread until after the war

which led directly to the development of the modern integral steel body in which all the sheet metal is stressed to withstand driving and suspension loads.

The first integral steel body for a small, popularly priced car appeared on the Opel Olympia launched at the Berlin Show in 1935. It was followed two years later by the 10 hp Vauxhall H model, both companies being linked by General Motors ownership. American capital provided the totally revised manufacturing facilities required by the new technology, but it was not until the rebirth of the European industry after World War 2 that other makers were able to adopt it as well. Ironically, in view of Opel and Vauxhall parentage, the separate chassis hung on in the USA for much longer, probably because steel was cheaper there, cars were much larger, excessive weight was not important and the public demanded high comfort and much lower levels of interior noise.

In the early stages of its development, the construction of an integral body was based on a platform of pressings, welded together in jigs which accurately located the attachment points for powertrain and suspension. Essentially, this platform consisted of front longitudinal members with mountings for the engine and suspension, side members acting as door sills, a transmission tunnel and rear seat pan braced by floor sections and the boot floor and spare-wheel housing. In some applications mechanical assemblies were mounted in subframes, or partial chassis structures to insulate the body from noise and vibration.

To achieve high production rates, the outer bodyshell was assembled simultaneously with the floorpan, to be joined later with the roof panel. Control of production differences or tolerances, achieved by precision jigs at each assembly stage, was a crucial factor in the success of the final mating together.

From rather overweight and understressed designs in the early days, the integral steel body evolved through the 1940s and 1950s to the efficient, relatively lightweight structures of the 1960s, as typified by the first Ford Cortina. This design employed new principles of low weight and cost to provide more car for less money

The 1962 Ford Cortina saved a considerable amount of body weight by optimizing each body panel to support part of the structural and suspension loads. Front and rear boxed members were joined only by thin side rails, a central transmission tunnel and the exterior sheet metal

Although pneumatic tyres on artillery wheels had arrived by 1902, chassis and suspension design on this Ford Model N still used back-to-back semi-elliptic leaf springs with a straight front axle

in the highly-competitive European market of 1962. Instead of employing the age-old technique of building in adequate strength from the beginning of the design, it was stressed to the maximum, tested in prototype form and reinforced only where there were weaknesses.

The Cortina broke away from using conventional side rails, and had braced panels forming torque boxes front and rear which were joined only by the sheet metal of the passenger compartment. Once the design had been refined, this construction proved both efficient and extremely good in absorbing crash energy in accidents. The weight saving it achieved meant that performance could be increased and, later, that fuel efficiency could be improved. It introduced some vibration problems in the early stages of development and led to early corrosion of some panels, where closed cavities were inadequately ventilated, but it was the beginning of a new era in body design, founded on a far more sophisticated structural technology than had been possible before.

The development of suspension systems

The need for a suspension system in a vehicle is not entirely caused by human susceptibility to the discomforts of vibration, although isolation from road-induced disturbances is certainly a highly desirable design aim. Maintaining contact between the wheels and the road surface is a far more fundamental and important reason, as the control and stability of the vehicle relies upon it totally. To provide isolation and insulation in the simplest way they could devise, early coach-builders developed the simple, semi-elliptic leaf spring, which was rolled at the ends to provide easy and flexible attachments.

The leaf spring is simply a flexible beam, supported at each end with a point load at its centre. Its main advantage is that its relatively wide but thin cross-section gives a high degree of stiffness in the horizontal plane, providing positive axle location, combined with flexibility in the vertical plane, for good isolation. But the internal stresses of a single-plane leaf (or a beam) supported in this way are far from uniform and much higher at the centre. To distribute them more evenly, the vertical section depth must taper in proportion to the distance from the load point. If this section was produced in solid material, it would be uniformly stressed but also very rigid. However, if secondary leaves are added instead, as a series of laminates following the same tapering section, flexibility and the ability to store energy are greatly improved. As each leaf deflects towards its elastic limit, it transfers additional loads to the next leaf in the stack, with a small degree of relative sliding motion between the two.

In the early days of suspension design, the attraction of the leaf spring was its suitability for manufacture in the primitive smithy, using existing skills and techniques. Easily malleable, mild steel could be formed to the required shape using only an open coke furnace, anvil and hammer, then tempered by quenching in water. It was hung from the chassis on flexible pivots and shackles and clamped to the axle with U-bolts and plates.

At the beginning of car development, the absorption properties required by the rigid tyres of the cartwheels made the double, semi-elliptic leaf spring the most commonly used suspension medium. Cars like the original 1886 Daimler had a pair at each corner, with the axle clamped below the lower set and the chassis clamped above the uppers. Other pioneering designers mounted the springs transversely so that a single-pivot steering system could be used for the front axle. Seventy years

It was the thrust to improve cornering power and handling in competitions that led to the dropped axle and upswept chassis frame, as on this 1914 Mercedes Grand Prix racer

CONTINENTAL

Massive ladder chassis, swept up over front and rear leaf springs, remained in vogue until the late 1930s. This Mercedes SSK would have been completed in this form before being shipped to a coachbuilder for its bodywork

DEUTSCHES MUSEUM MÜNCHEN

later, transverse leaf springs were still in use in independent suspension systems on cars like the Chevrolet Corvette and AC Ace.

The use of the coil spring as the primary suspension medium did not occur until the more general trend towards independent front suspension after World War 2, and even then it was used in conjunction with leaf springs at the rear. In addition to being more efficient, the coil spring offered the advantage of an almost total lack of internal friction (caused by the sliding action between the leaves of a leaf spring design), which allowed the damping function to be confined to a more precisely controlled shock absorber unit.

Leaf springs were easy to incorporate in the early flat, ladder-type chassis, since they required widely-spaced attachments in the same horizontal plane. Coil springs, although much more efficient in their ability to store energy and their range of deflection, need to react vertically against their mountings. The concept of the coil spring is inseparable from the integral body or the three-dimensional, space-frame chassis.

Suspension design today has polarized towards coil springs in a variety of forms because of these design criteria and because they can be manufactured relatively cheaply in large quantities with consistent characteristics. In the modern, highly-efficient bodyshell they also feed road loads into well-spaced and stiff areas of the structure.

In the history of suspension development, there has probably been more progress in the understanding of the effects of wheel geometry and kinematic changes than in any other related field. Only in the 1980s have the ideas of self-aligning wheels that can compensate for any imbalance in the braking system or tyre drag, of negative roll in corners or of the fine-tuning of handling through rear-wheel steering started to become feasible and viable for popular cars.

In the following chapters the principles which underlie suspension design, the methods used throughout the automotive industry and the relationships between other chassis features are explained and discussed in detail.

Designed by Dr Ferdinand Porsche in 1934, the chassis for the 200 mph Auto Union Type C racing car was based on twin-tubes and compact, torsion-bar springing

2 Introduction to the science

The definition of the chassis and its function as the main supporting medium for the suspension have been outlined in the preceding chapter. More detail on modern methods of optimizing the stress-bearing elements of integral bodyshells will be found in Chapter 6. Before embarking on any explanation or description of suspension systems and the complexities of vehicle dynamics, it is necessary to understand the terminology used. Several key definitions are introduced here, with others being added as footnotes where appropriate.

By definition, a suspension system supports a load from above, which is exactly how the early engineers and mechanics saw the role of the springing medium (their primitive carriages were hung from their supports) and why the apparent misnomer came into general use. In reality, the mechanisms that isolate the occupants of a modern car from road disturbances are *compression* systems.

Springs are the flexible elements in the suspension system which are able to store the energy applied to them in the form of loads and deflections. Different types of spring have different qualities in this respect (see Chapters 3 and 4). In most countries, spring flexibility is measured as the rate at which loads cause spring deflection, and newtons per metre are typical SI spring rate units (France is one of the exceptions where metre/newton is sometimes used).

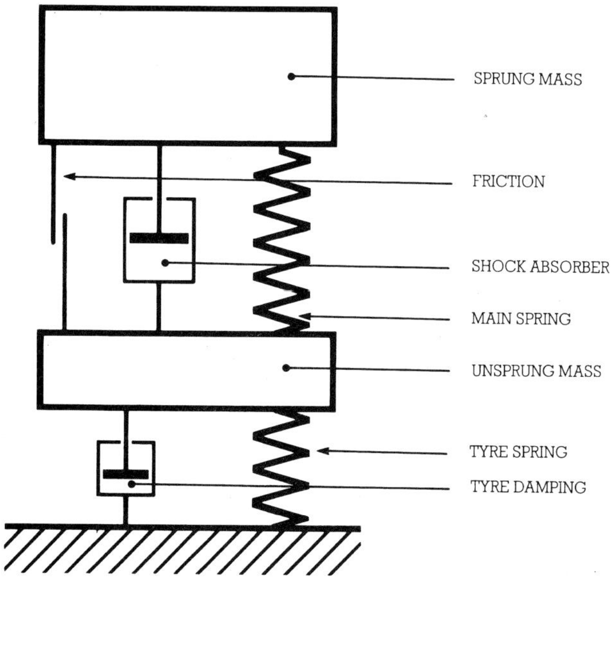

Each wheel can be simplified to a system represented like this, with the sprung and unsprung masses supported by the main spring and tyre as shown. Damping and friction are additional elements to be considered separately

Spring energy is converted to heat and dissipated partly by the friction in the system and mostly by the dampers, more often and erroneously called shock absorbers. Dampers usually take the form of pistons working in cylinders filled with hydraulic fluid. They exert a force which is proportional to the square of the piston velocity (see Chapter 3). The damper's function is to restrain undesirable bounce characteristics of the sprung vehicle mass and to prevent the wheel assembly from losing contact with the ground by being excited at its natural vibration frequency.

The remaining mechanical elements in a suspension system locate the wheel assemblies and control the geometry of their movement. Some of these elements are simple links, others may be compound, multi-role members like the relatively thin, transverse torsion bars used to stabilize the vehicle in corners by restricting roll (the stabilizer or anti-roll bar).

In engineering terms, 'mechanics' describes the science that deals with the action of forces on objects or structures. It forms the basis for all mechanical design—the analysis of loads, functions and motion. Under the influence of forces, all objects may be either in motion or at rest, relative to some other defined system or reference point. The area of mechanics that deals with motion is called 'dynamics' (statics are concerned with forces in equilibrium and kinematics refer to the theory of motion without reference to the forces involved).

In the simplest case of a particle in motion, its rate of movement is known as its speed, without its direction of movement being specified. Speed has units of length or distance divided by time. Quantities which combine magnitude with direction

In addition to moving vertically over bumps and horizontally along the road, a vehicle has many additional degrees of freedom, usually simplified to act about the three primary axes (X, Y and Z)

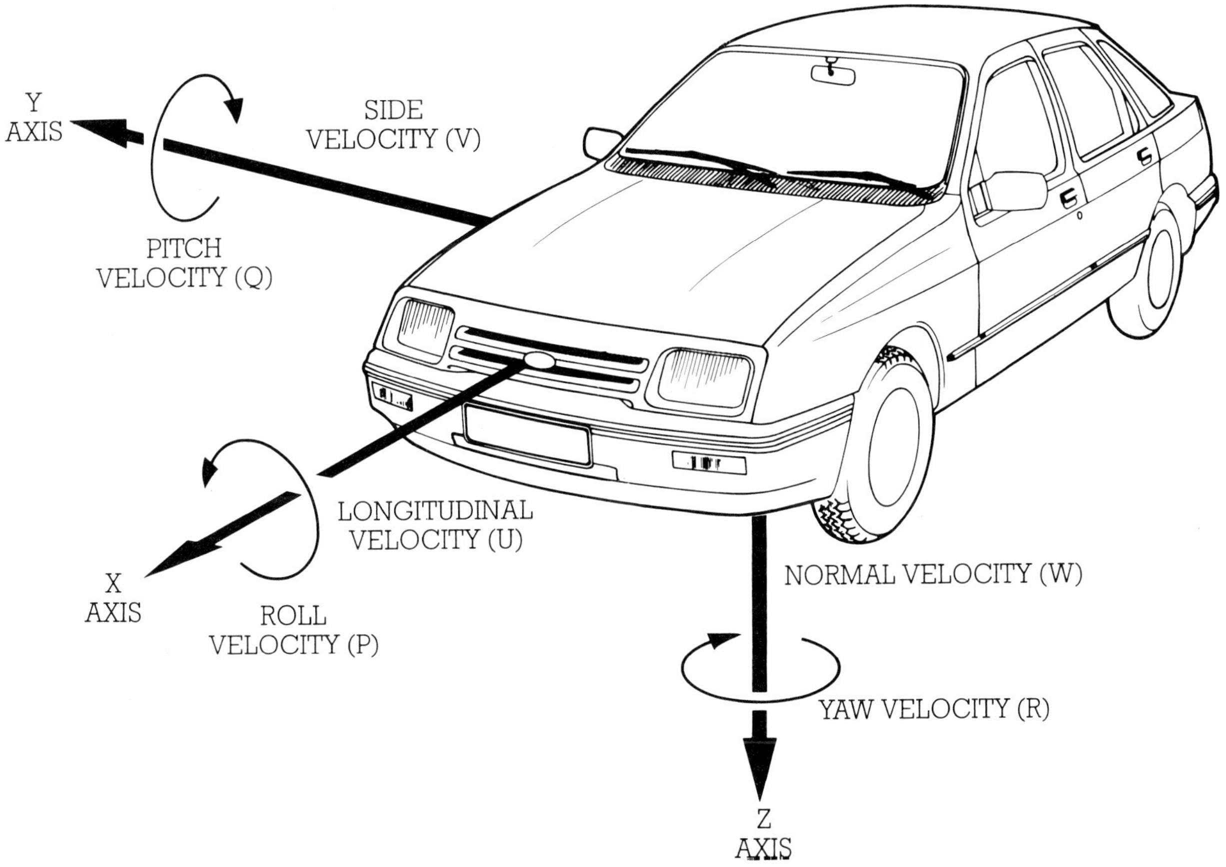

are known as vectors, the most commonly used vector in vehicle terminology being velocity (which is therefore clearly differentiated from speed). The rate of change of velocity for an object is its acceleration, and its velocity multiplied by its mass is its momentum. The actions of forces and momentum are governed by several laws of motion which explain the behaviour of everything we see moving around us.

Vehicle dynamics

The primary function of a vehicle's suspension system is to isolate the structure (and the occupants) from shocks and vibrations generated by the road surface. It must do this while maintaining stability, steering control and overall handling precision. Therefore, each wheel is connected to the chassis or integral body of a car by a system of links, which control the kinematics of the suspension movement, and a system of springs and dampers, which provide flexible but restrained wheel movement. The diameter of the tyre, the flexibility of its sidewall (which acts as a secondary spring), the size of the contact patch between the tyre and the road, and the (unsprung) weight of the wheel, tyre and axle assembly all affect the magnitude of the shock transmitted to the suspension unit. The spring rate or stiffness, the damping effects of the shock absorber and the ratio of sprung to unsprung masses affect the amplitude of the wheel motion and the load transmitted to the body structure through the spring seats and suspension attachment points.

Apart from the forward motion of the vehicle and any cornering forces generated by steering action, the total vehicle system is subjected to many different degrees of free movement, acceleration and vibration, simplified in most analyses to act about three axes. In addition to moving vertically on its springs, a vehicle is free to yaw about a vertical axis, to pitch about a lateral horizontal axis and to roll about a longitudinal axis passing through the front and rear roll centres (provided all wheels remain in contact with the road surface).

The interaction of these movements, each with its own velocity, acceleration and frequency, makes a sprung vehicle one of the most complex systems in the field of dynamics. With at least 17 different degrees of freedom, analysis of what takes place during a complex road manoeuvre (which often involves coincidental roll, pitch and yaw) cannot be undertaken adequately without computer aids, especially when there are additional variables to be considered, such as body flexibility, suspension-link deflections, compliance in the attachment bushes, and geometry changes caused by torque inputs to the wheels and by suspension movement.

Most modern cars use independent, coil-spring suspension front and rear which has been developed to provide dynamic characteristics that combine good ride qualities with outstanding handling and driver control. Earlier car designs, many of today's vans and most trucks have rigid axles, the last two for sound engineering reasons. Over the years, a number of other suspension designs have been tried, usually without success in the long term. However, the way ahead seems to lie with a combination of variable-rate air spring and hydraulic control under the influence of an on-board microcomputer (see Chapter 9).

The primary function of a suspension system has changed little, however, even if the understanding of its behaviour and the quality of its ride comfort have both made quantum leaps ahead. A suspension spring is still required to absorb the shock loadings produced by bumps in the road surface, which it does by being compressed initially and subsequently expanded against the total mass of the car. The rate at which energy is stored depends on the spring rate, or stiffness, and the rate at which it releases this stored energy is controlled by the natural frequency of the sprung system, typically 1 to 1.5 hertz (or cycles per second).

Effective damping is essential to prevent spring deflections running out of control, particularly through resonance of the system around its natural frequency, and to control wheel and body movements. In ride terms, the subtlety of suspension design depends to a large extent on the balance between spring stiffness and damping effect, interacting with the amount of wheel travel available and the means of restricting it. Progressive springing can be achieved either by inherent spring characteristics (such as in coils wound from tapered-section wire) or by incorporating flexible bump stops, or snubbers, in the system.

Major constraints on this aspect of suspension design concern the ability of the vehicle to carry loads without reaching the wheel travel limits under normal running conditions. In this respect, there is much to be gained from load-levelling control systems, although their complication adds considerably to the basic component costs.

If each wheel could be considered as a separate system, suspension design would be a much simpler process. But all the springs are linked together by the structure of the chassis or body and all the wheels by the reaction of the ground, which makes their interactions complex and their characteristics interdependent. A car is free to bounce equally at both ends at once, to pitch about the front or rear axles, to roll about a longitudinal axis and to combine any or all of these effects. How the static masses are distributed, how the wheels are connected and how the spring rates are balanced make a fundamental contribution to the behaviour of the vehicle under dynamic conditions.

Suspension geometry

Kinematic variations in wheel angle and movement are generally described by their geometry, and the science of wheel control has become as complex as ride-quality development. To understand the various systems in detail, it is necessary to define the variables first.

Toe setting This is the small angle between each wheel and the longitudinal axis of the car. It is usually measured statically by taking the difference in the distance between the front and rear edges of left and right wheel rims at centreline level. Under dynamic conditions, toe settings often change due to longitudinal and lateral movement of the suspension and body components.

Castor angle and mechanical trail Usually, a steered wheel is arranged to trail by a small angle, so that forward movement gives a stabilizing effect. The angle of trail is known as the degree of castor. Most modern cars use the self-aligning torque of the front tyres and the front weight bias to provide self-centring of the steering system and true running, so castor angles are modest and typically 1 or 2 degrees.

Camber angle This is the angle between the plane of a wheel and the vertical. It is described as positive when the top of the wheel leans out and negative when the top leans in. Camber angles usually change slightly under the geometry influences of wheel travel. As camber angle also affects the pressure distribution of the tyre footprint on the road, it plays an important part in the balance of cornering power and handling characteristics.

The more positive the camber angle, the greater the slip angle for a given cornering force and the slower the onset of complete breakaway. The more negative the camber angle, the higher the ultimate cornering force, but total

Each geometric angle of the wheel is defined by terms which describe its inclination to the vertical axis through the hub centre. The offset of the contact patch from the steering axis is measured both laterally (as king pin offset, wheel offset or scrub radius) and longitudinally (as castor offset)

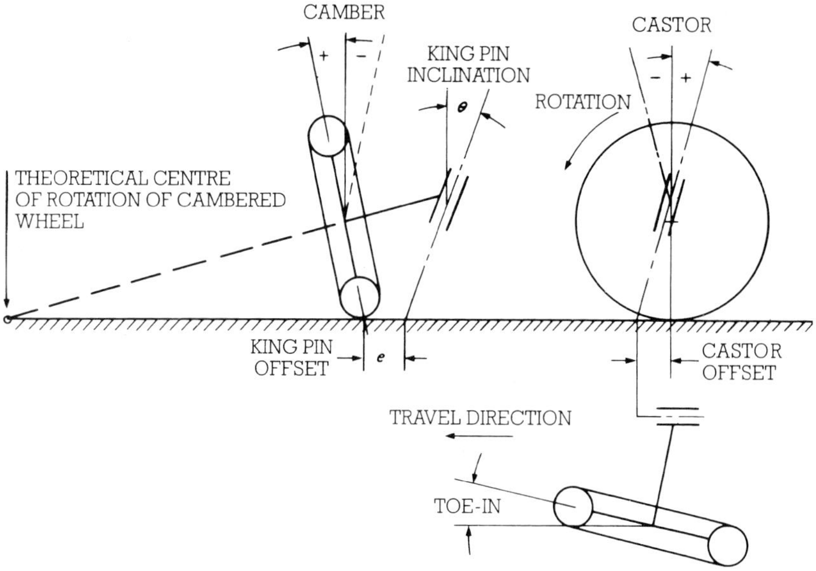

Negative wheel offset provides stability by counteracting any uneven forces automatically

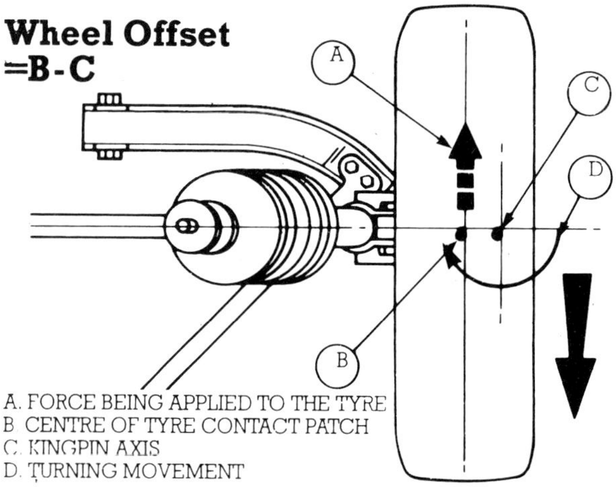

breakaway becomes more sudden and rapid. For maximum traction, as well as cornering power, high-performance cars with wide tyres need to run with low camber angles on their driving wheels and as little change in the angle as the suspension installation allows.

Wheel offset (or scrub radius) The lateral distance between the point where the swivel axis of a steered wheel meets the ground and centre of the tyre footprint is known as the offset. If the axis passes inside the footprint centre, the offset is positive, and if it passes outside, offset is negative. If it coincides, the system is termed centrepoint. Negative offset provides stability advantages in the event of uneven braking torque or a tyre deflation, but is often claimed to reduce steering

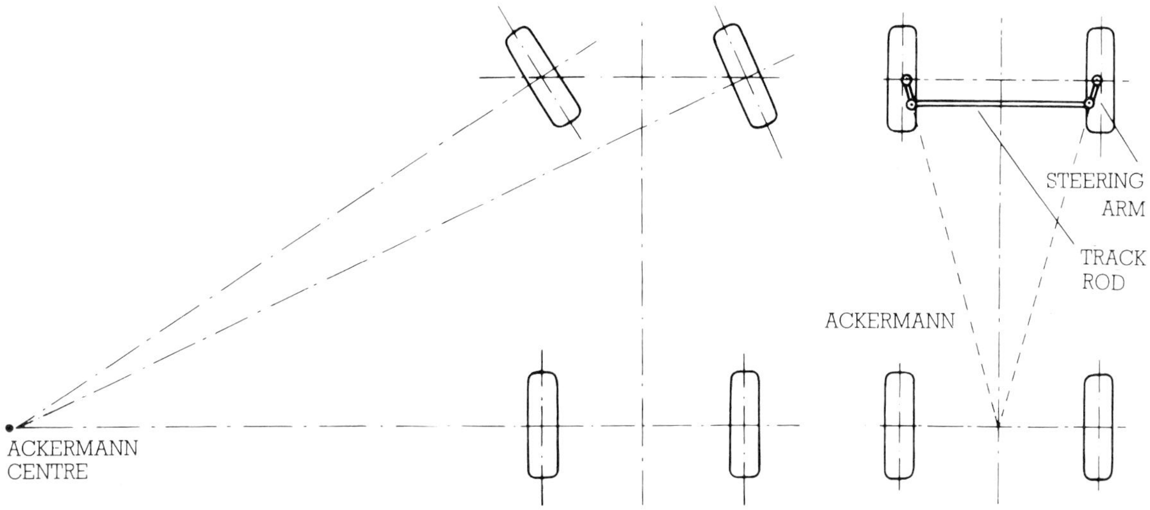

Ackermann effect allows the inner steered wheel in a turn to cover a tighter radius than the outer by angling the steering arms towards each other slightly, to intersect at the centre of the rear axle

feel. Positive offset is claimed to generate a greater self-centring action but increased steering effort, although opinions on these factors often differ.

Ackermann effect This is the degree by which the inner wheel in a turn is arranged to move through a tighter radius than the outer wheel. Full Ackermann effect is provided by setting the steering arms at a slight angle so their projected lines meet at the rear axle and all four wheels run around a common centre in bends. The importance of full Ackermann geometry has declined with advancing tyre technology, although Ford investigations into the optimum balance between offset and Ackermann effects have produced a system, first introduced on the Sierra, which combines low steering effort with excellent feel.

Roll centres and roll axis Under lateral cornering forces, the body will roll on its springs about centres at both ends of the car. The line joining these centres is the roll axis, and its inclination to the horizontal plane plays an important part in handling characteristics. Determining the roll centres for a MacPherson strut suspension or an independent, trailing-arm system involves some geometrical plotting (they are usually just above ground level, but both systems allow them to be adjusted through the design geometry of the various links involved). On a MacPherson strut system, the roll centre moves vertically with bounce and rebound, and from side to side with body roll. To restrict roll, left and right suspension arms are often linked by a slim, torsion-bar spring, often called a stabilizer.

Anti-dive and anti-squat To resist front-end dip under heavy braking, or rear-end squat under acceleration, suspension pivots are usually set at an angle to provide upward reactions automatically in response to high wheel-torque inputs. Some movement is preferred by most drivers to improve braking feel, but it is well controlled in most modern cars.

Torque steer With front-wheel-drive cars, there can be an interaction between the drive system and the suspension geometry which causes directional influences on

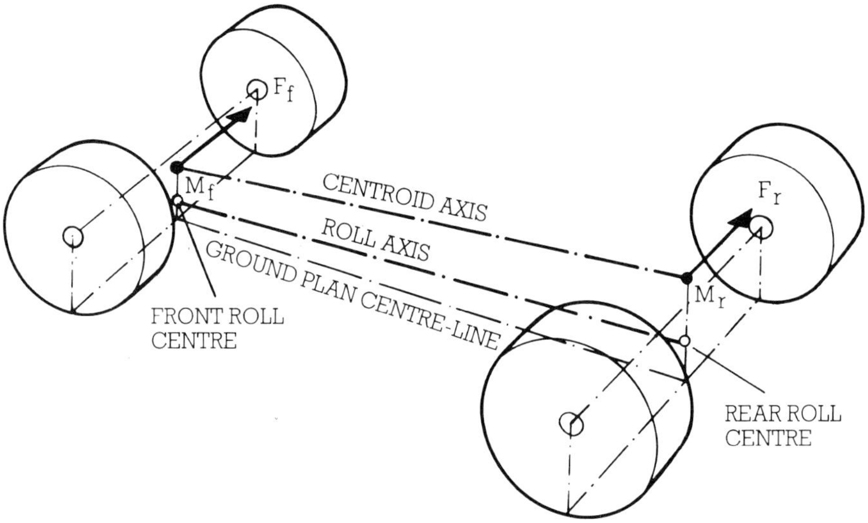

The axis between the front and rear roll centres is the roll axis. Its inclination plays an important part in handling characteristics

Anti-dive effects under braking can be counteracted by using a leading-arm geometry that generates an upward reaction in response to high wheel torques. Double wishbone layouts or struts can be given similar characteristics by introducing opposed inclination between their inner pivot axes

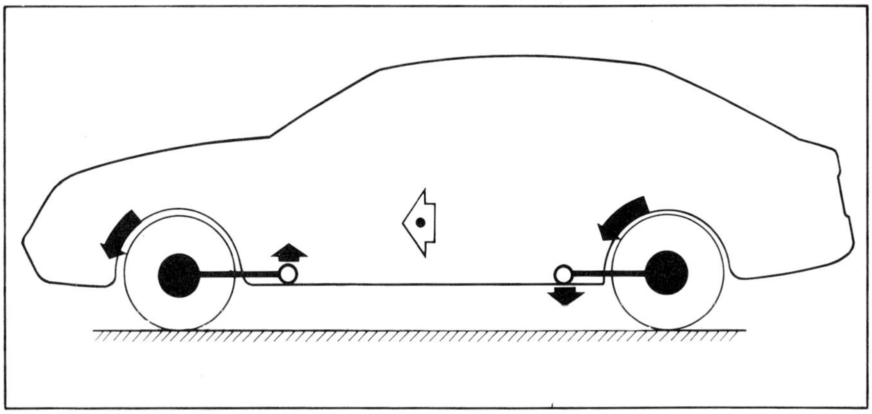

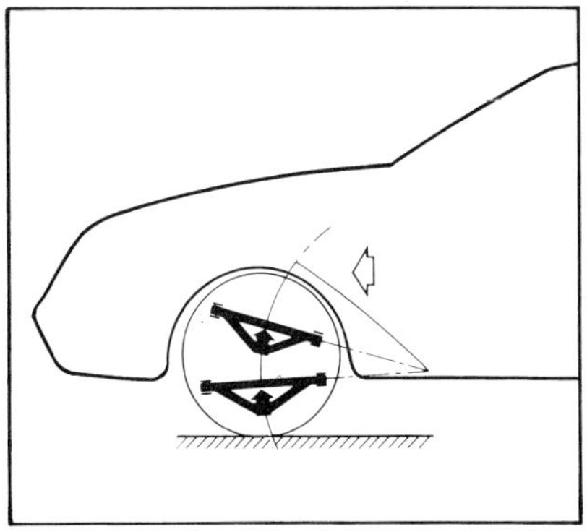

the steering, known as torque-steer effects. They are detected as an easily-countered pull on the wheel when accelerating hard, usually associated with a tendency for the car to steer the opposite way when the engine power is suddenly reduced.

Torque steer is one of the most difficult design challenges with front-wheel drive, and the difficulty increases with higher power levels. Common causes of torque steer include unequal drive-shaft stiffness, unequal changes in drive-shaft angles on opposite sides of the car, or a conflict between drive shaft, suspension or steering geometry.

Roll steer There are several ways in which the effects of roll can cause either the front or rear wheels to steer the car slightly, and this characteristic may be used for fine tuning of its handling. In some cases, the toe setting may change as the outer wheel becomes more loaded in a turn, or the outer wheel may creep forward slightly (as the inner wheel creeps backwards) to create a steering effect. Rear roll steer is often used to counter front understeer, either through the primary geometry of the system or by secondary loading of the compliance bushes. Too much roll steer, however, can lead to excessive tyre wear and disturbing bump-steer effects on long-wave, single wheel movements such as those experienced on motorways.

3 Spring and damper theory

In its simplest schematic form, the suspension of a vehicle can be considered as a basic two-mass, two-spring system (as shown in the diagram below). The primary mass is the bodyshell with all its mechanical components, trim, equipment, occupants and luggage. It is supported by the primary springing medium which, today, is most often a set of helical, constant-rate coil springs. The secondary mass is made up of all the individual masses of the unsprung components of the suspension, drive, braking and steering systems. They are isolated from the road surface by the tyres which act as secondary springs.

In reality, the primary spring has its own secondary elements, such as insulating pads, bump stops and the inherent flexibility in the suspension components, but their effects can be excluded from the analysis at this stage. The dampers also provide a degree of minor secondary springing which can be ignored as well. Initially, we shall consider the bodyshell to be a perfectly rigid structure, although

If the sprung mass is considered to be supported by a primary spring and damper, and the unsprung mass by a secondary system that includes the tyre characteristics as shown schematically here, the overall rate for the compound system can be calculated

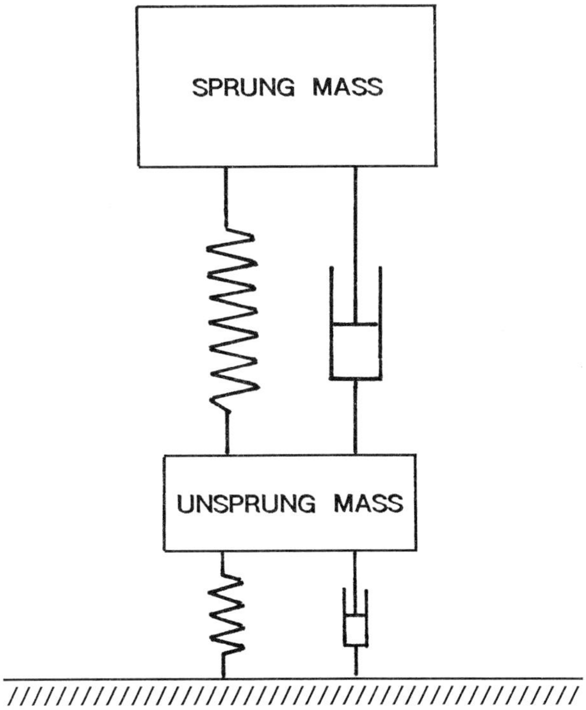

its flexibility cannot be disregarded if calculated predictions are to be truly representative of actual effects.

The overall rate of a compound spring system can be determined in the same way as calculating electrical resistances—straight addition for those in parallel ($R_{TOTAL} = R_1 + R_2$), and by adding the reciprocals of those in series:

$$\frac{1}{R_{TOTAL}} = \frac{1}{R_1} + \frac{1}{R_2}$$

Therefore, the theoretical straight-line relationship between load and deflection of a linear-rate spring is distorted into a curve by the other flexible elements in the system. To prevent contact between suspension elements or the spring coils at full travel, bump stops are added which distort the load/deflection curve even further.

The energy stored by the spring is represented by the area under the load/deflection curve, whatever shape it takes. Different types of spring have different capacities to store energy at the same stress levels. Their performance characteristics and relative fatigue life depend significantly on this capacity value. In general, leaf springs can store much less energy than torsion bars or coil springs stressed to the same level, so are about four times heavier for any given installation. Usually, some of this weight can be offset by the elimination of locating links.

Spring media other than steel have completely different characteristics. Rubber,

In reality, a suspension system comprises many sprung and unsprung elements that include tyres, bushes, locating links and anti-roll bars

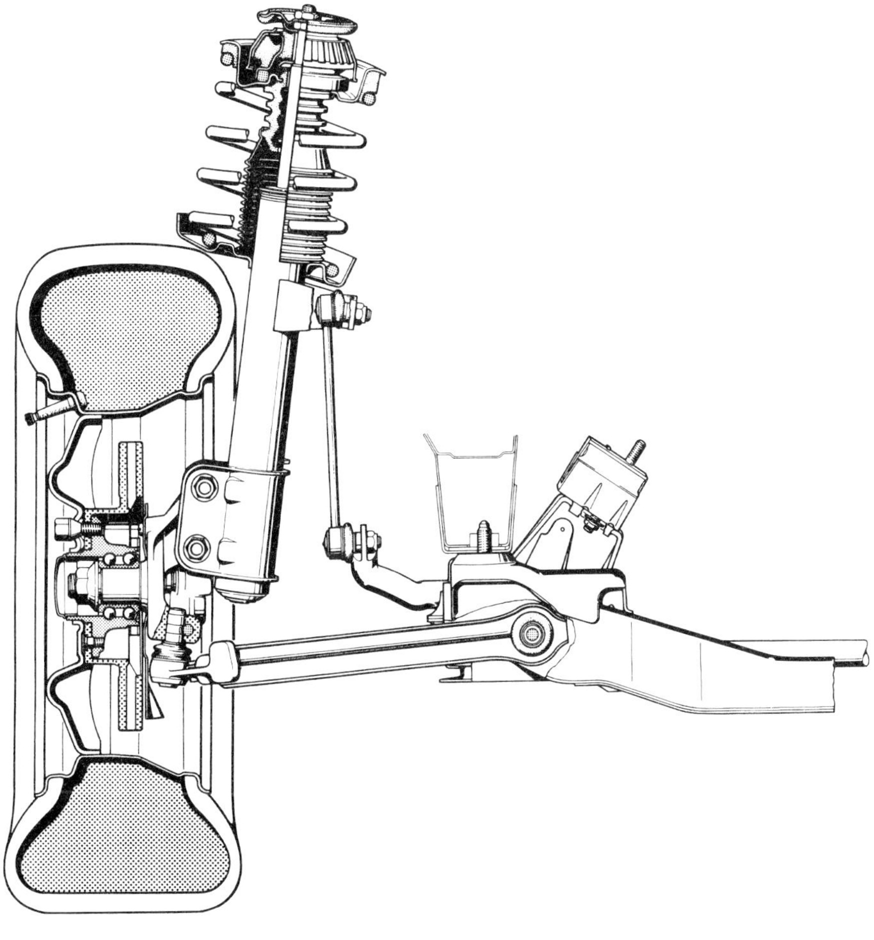

for example, even when used in conjunction with an inert gas or hydraulic fluid (see Chapters 4 and 9), has a much higher capacity to store energy than steel, as does a pure gas. Although the actual levels are difficult to calculate precisely, because of all the possible variations in design and configuration, the following table gives typical examples.

Typical energy capacities	*(Nm/kg)*
Leaf spring	50
Torsion bar	200
Coil spring	200
Rubber cone	375
Hydrolastic rubber	750
Gas springs	1500*

*This figure is highly dependent on the materials used and the ancillary equipment needed in the form of compressors, reservoirs, sensors, operating valves etc.

In dynamic terms, the maximum load likely to be fed into a vehicle road spring will be about twice its loading in the static laden condition. If, for example, a spring has a stiffness rate of about 30 kN/m, it will be compressed about 3.15 in. (8 cm) by a typical static load of 547 lb (248 kg) (generating a force of 2400 N) but will be displaced a further 6.3 in. (16 cm) by road inputs. Therefore, to reduce the clearance needed for suspension movement to below 6.3 in., some form of supplementary springing is required. This can take the form of rubber or polycellular cones which distort progressively or variable-rate springs wound from tapered wire. In a typical installation, bump rubbers would come into operation at about 4 in. (10 cm) dynamic travel and restrict the total movement to about 5 in. (12.7 cm). The mean linear stiffness needed to do this would be 37.8 kN/m and, therefore, the bump-stop stiffness needs to be only 7.8 kN/m (from the parallel rule quoted earlier).

Coil springs, in effect, act as torsion bars, but conveniently are wrapped up into a more efficient package which is easier to install, especially in the confines of a car's wheel-arch area. Because the bar is twisted into a helical coil, shear stresses are greater than for a straight torsion bar, but are easily allowed for with modern manufacturing and heat-treatment methods.

Spring oscillations after a single bump deflection decay with simple harmonic motion at the natural frequency of the total system

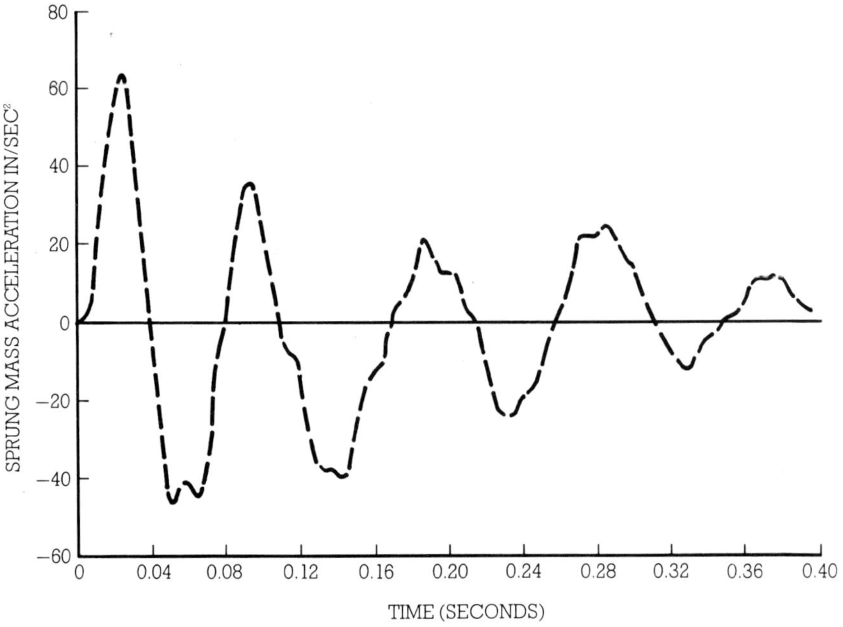

Natural spring frequencies

Any mass supported by a spring and displaced from its position of equilibrium will vibrate with simple harmonic motion* at a constant, natural frequency determined by the amplitude of the initial displacement, the stiffness of the spring and the mass. The periodic time (time for one complete cycle, or the reciprocal of the frequency) is equal to 2π multiplied by the square root of the mass, divided by the stiffness. This frequency is a powerful factor in the behaviour of a suspension system on the road, for it determines the rate at which the energy stored by a compressed spring is fed back into the body structure—the residual effects of crossing a bump or pothole.

However, the road wheel and its suspension system are not a simple mass on a linear spring. As explained already, they form a compound system in which the relationships between the sprung and the unsprung masses, and between the primary and secondary springs (the tyres) are critical. Such a compound system can vibrate in two distinctly different modes. The sprung mass of the body structure can either be in phase with the unsprung wheel and suspension masses or out of phase with them. In the former case, the forces involved work together, and in the latter, they oppose each other.

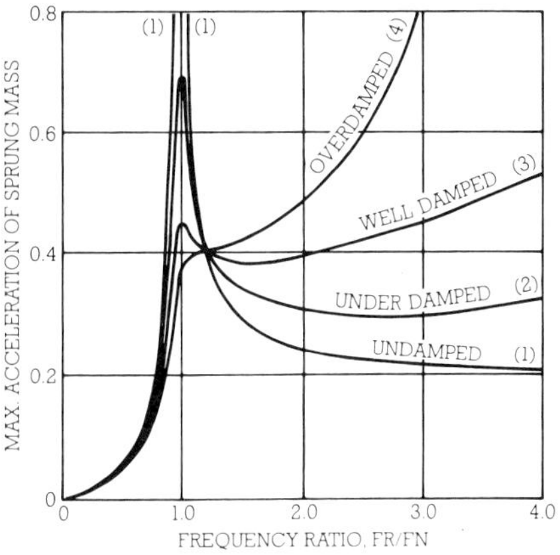

Optimum damping is a compromise which becomes critical when the sprung and unsprung mass frequencies are of similar values. In the undamped case shown here (1), a resonance occurs at a frequency ratio of 1.0, generating very high accelerations of the sprung mass. In the overdamped case (4), the high frequency ratios are harder to control

If this type of compound system is represented by an equivalent layout with the two elements considered separately, the true natural frequency of the sprung mass can be determined. Then this value can be fed into the compound system to calculate the unsprung natural frequency as well. By traditional methods, these values were always derived from successively closer approximations, but now all that tedious work can be avoided with modern computer programmes.

Newton's second law of motion (force equals mass multiplied by acceleration) can be applied to the unsprung masses of a car's suspension, which act upon the spring seats to generate an acceleration in the sprung mass of the body. Assuming no energy is lost in the system, the accelerations of the two parts of the system will be inversely proportional to their relative masses. The less the unsprung mass is relative to the sprung mass, the less will be the effect of wheel inputs on the structure.

*Simple harmonic motion is a type of vibration where the acceleration (and hence the restoring force) is proportional to the displacement of the vibrating mass from its position of rest. It can be represented by projecting the uniform movement of a point around the circumference of a circle on to a diameter.

Exciting the wheels at the natural frequency of the system on undulations or washboard surfaces can cause them to resonate and lose contact with the road surface. It can also cause the body structure to resonate on its springs at the same natural frequency with very unpleasant effects on the quality of the ride. Any input which excites this frequency in a totally undamped system would continue as a resonance forever. In the real world, however, there is friction damping, which exerts a constant force opposing any movement, air resistance damping of the moving parts, which can reach velocities as high as 16 ft (5 m)/sec (see Chapter 10) and is proportional to their linear velocity squared, and the viscous damping designed into the system through the hydraulic shock absorbers. This damping force is directly proportional to the velocity difference between the sprung and unsprung masses.

The effects of a two-axle vehicle crossing a single bump are compounded by the interconnection of the two spring systems through the chassis structure

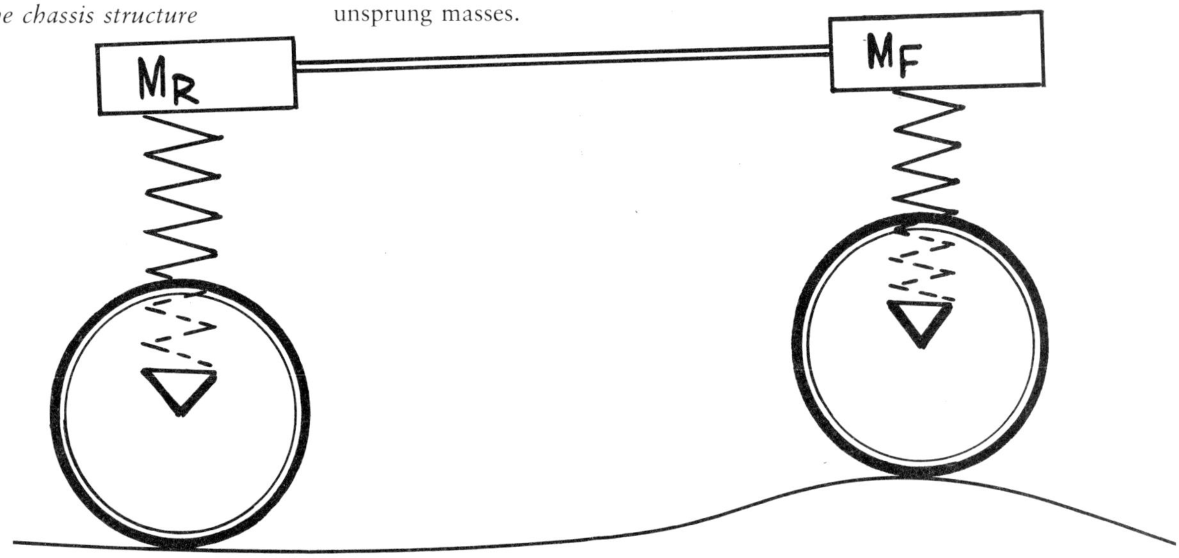

Critical damping effects

As dampers provide a resistance to movement between the sprung and unsprung masses, they also create an additional source of load inputs to the body structure from road shocks. Therefore, there are limits to the amount of damping that is acceptable for a comfortable ride.

In theory, at least, a vehicle would be most comfortable if the dampers were designed to provide minimum resistance to suspension movement on bumps and a far greater resistance on rebound. Although distinctly different bump and rebound settings are incorporated into most dampers in practice, their values are always much closer to each other than might seem best because of the inevitable interaction between the two directions of wheel travel.

For convenience, damping forces are often considered in relation to their critical values*. This critical damping force can be calculated simply by multiplying the mass by twice the angular velocity of the simple harmonic motion involved (also equal to twice the square root of the mass multiplied by the spring rate). In most applications, critical damping is far too severe, and the best damping for comfort and adequate wheel control falls somewhere between 15 and 50 per cent of the critical value, depending on the friction inherent in the system and the ride qualities required.

*Critical damping is the minimum damping effect required to prevent an oscillation passing the static position after a disturbance of the sprung mass.

The typical function of the damper in a modern car is to reduce the amplitude of the deflection over a bump to about a fifth of its initial value within two cycle oscillations at the natural frequency of the system.

Interconnected suspensions like this Morris 1100 Hydrolastic system, introduced in 1962, represented ambitious attempts to counter the natural forces at work

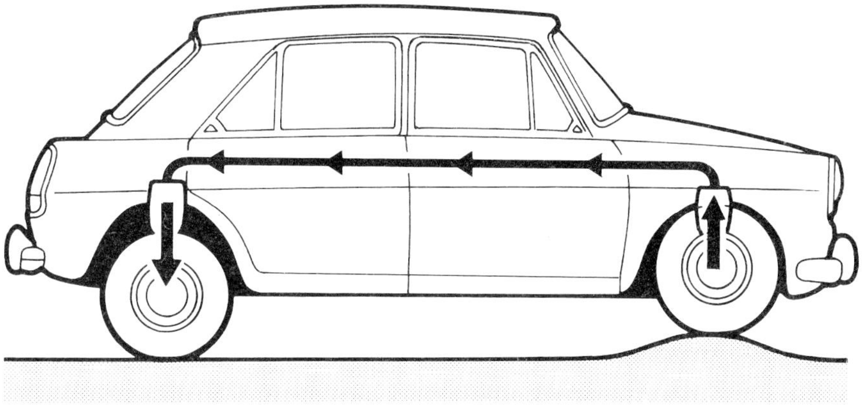

Interaction between front and rear

A conventional, sprung vehicle running on wheels is not a simple mass supported by a compound spring system. It is, in effect, a beam supported at both ends and free to pitch when considered in profile, and it acts as an independent beam free to roll when considered in the lateral plane. In total, it is a highly complex system with many different degrees of freedom (see Chapter 2).

For the purposes of analysis only, the motion caused about the two main axes of pitch and roll can be considered separately. In practice, the effects are interrelated and the system must be considered as a whole for representative results to be obtained by prediction. That process involves hundreds, or even thousands, of simultaneous equations which can be solved only by computers.

To understand the behaviour of a vehicle about its pitch axis, it is necessary to make a few assumptions and extrapolate to the extreme conditions. Starting with mass distribution as a controlling factor, the reader should imagine the limiting circumstances of a rigid, unsprung vehicle with a long wheelbase and 99 per cent of its mass concentrated on, say, the rear axle rather like a dragster. In this case, the centre of gravity is very close to the rear axle, and the effects of front-wheel deflection over a ridge in the road would apply a very high leverage to the mass, the resultant pitching being very difficult to damp out. The car would rock about the rear wheels.

If the mass is moved to the centre of the longitudinal beam joining front and rear wheels, the effects are much more balanced and the motion over bumps is much closer to the ideal case of pure bounce. The phasing of the inputs between front and rear would be delayed by the time interval between the front and rear wheels hitting the ridge, which is dependent on the forward speed.

A distributed rather than a concentrated mass also has a polar moment of inertia (see Chapter 2) determined by the nature of its mass distribution. This has a considerable effect on the residual pitching motion and the speed with which it can be damped out. The inertia of a sprung mass will always resist pitching, so the further the mass is concentrated from the centre of the beam, the greater the resistance to pitch. Unfortunately for the car designer, high polar moments of inertia have a detrimental effect on handling (see Chapter 5).

Suspension systems are subjected to three distinctly different kinds of dynamic forces, derived from the road surface (top), the vehicle's own reactions to acceleration and braking (centre), and cornering effects (bottom)

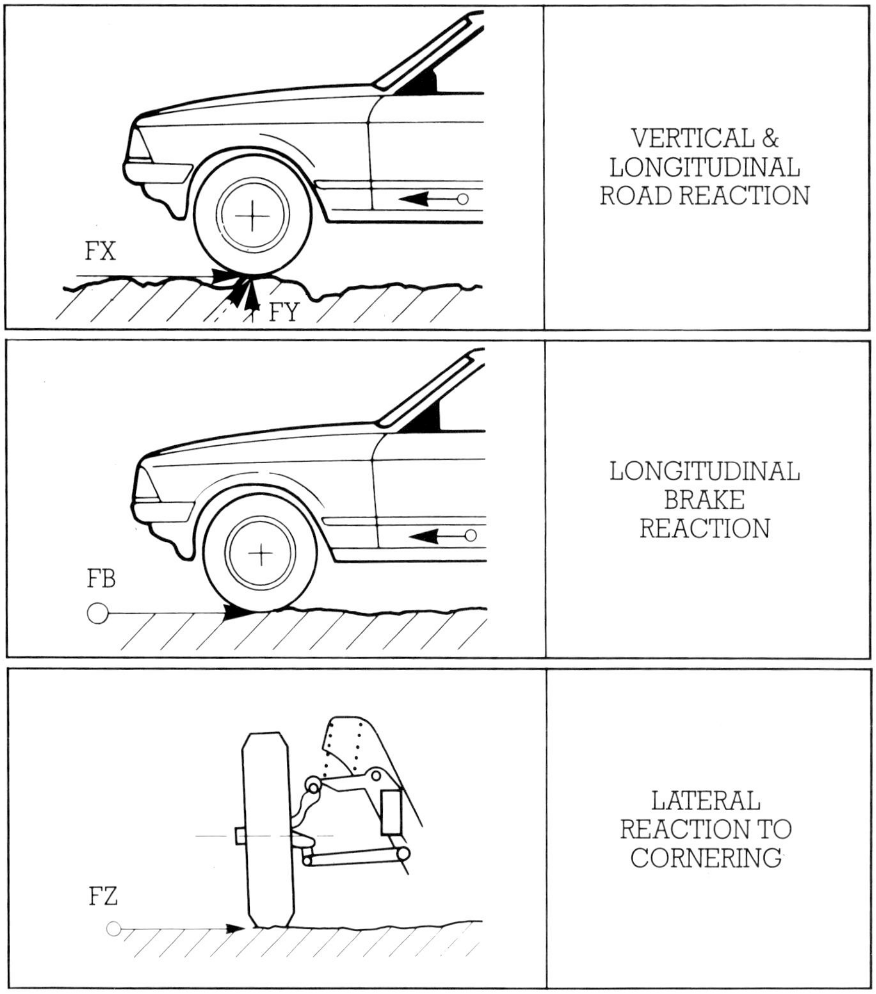

Other factors which influence pitch characteristics are the wheelbase and the relative frequencies of the front and rear springs. Pitching moments depend upon the pitch stiffness and the wheelbase, which sets the leverage effect of front and rear inputs about the centre of gravity. A short-wheelbase car is very stiff in pitch because the lengths of the lever arms are short.

If we move on to the more typical case of a sprung car with an average wheelbase and close to 50:50 mass distribution, the effects of a single bump or dip can be visualized more clearly. For this purpose, it should also be assumed that the disturbance is either consistent across the road (both wheels at each end are displaced together) or the analysis applies only to one side of the vehicle (diagonal inputs will be considered later).

As the front wheels hit the bump, they are displaced upwards against the spring forces, and a reduced input to their mounts lifts the body against the gravitational force of the front-axle mass. A similar effect occurs at the rear when the back wheels hit the same bump after a time interval dependent on the speed and wheelbase. Both displacements excite some pitch movement, and the rear input may work either with or against the initial disturbance, depending on the pitch frequency and the delay between the inputs.

If the suspension movement at each end was to be critically damped (damped sufficiently strongly for the oscillations to die out after one complete cycle), pitching motion could be eliminated rapidly, but the vertical ride qualities would be extremely firm. Instead, a compromise is struck between vertical stiffness and the pitch resistance for balanced standards of comfort. The relative suspension frequencies at front and rear also decide the amount of initial pitch reaction.

In theory, interconnection between front and rear systems is a simple way to relieve pitching problems, but usually it is a difficult concept to apply at a reasonable cost. If the same spring is shared between front and rear suspensions on each side (see Chapter 9), any pitching moment caused by deflection of the front suspension is immediately countered by an equal and opposite moment at the rear, cancelling out the disturbance. However, since the centre of gravity is never on the line joining front and rear suspension pivots, it causes additional moments, usually described as dynamic weight transfer. These are not resisted by freely-mounted, linked springs, so the interconnection must be compromised to reduce the excessive squat and dive that would otherwise cause the car to hit its bump stops at the rear when moving off and at the front under braking.

To calculate some of the pitch effects, it is necessary to apply a theory of machines known as 'conjugate points'. This is based on the assumption that a rigid body mounted on springs at both ends oscillates about predetermined virtual centres displaced from the spring centre of the total system (the point of equilibrium through which vertical forces produce only vertical deflections), biased from the mid-point by the ratio of the front and rear spring rates. If, for example, the front springs are twice as stiff as those at the rear, the spring centre will be a third of the distance between the two springs measured from the front.

The analysis shows that the product of the two spring lever arms ($\frac{1}{3}$multiplied by $\frac{2}{3}$of the wheelbase in the example) will also equal the product of the distance ratio between the displacements to the two virtual centres from the spring centre. Once the positions of these centres, also called the elastic conjugate points, have been

This typical plot of damper force against stroking velocity shows how the low-speed performance can differ significantly from faster movements

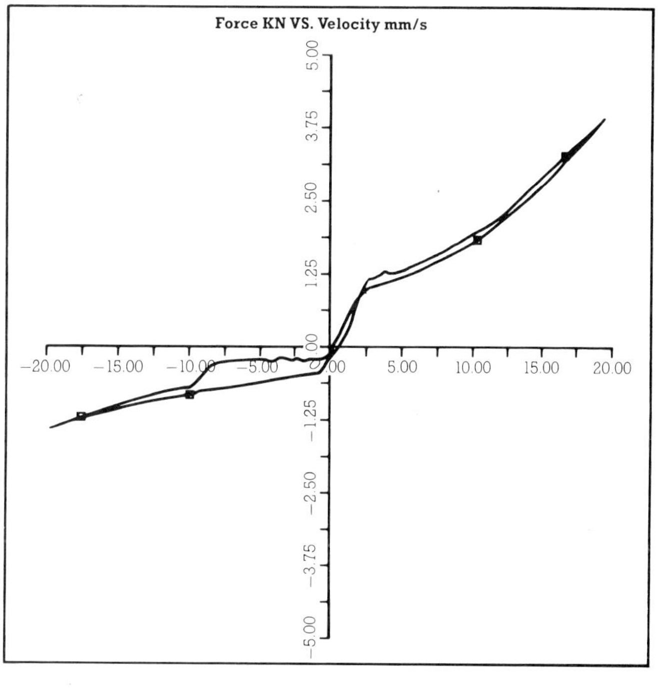

determined by calculation on a simplified model or test measurements on a complete vehicle, the same logic can be applied to the dynamic movement using inertia relationships instead of mass. There is only one place where each elastic conjugate point coincides with its dynamic conjugate point and that sets the pitch behaviour.

So although it might seem relatively obvious to select front and rear spring rates simply to satisfy just the loading and free travel conditions applying at each end of the car, the ratio between these rates has a significant effect on the pitch characteristics. As, by definition, no motion is generated at one conjugate point by forces applied to the other, many suspension systems are designed to have their conjugate points coincident with the axle centres, to reduce the tendency to pitch. The difficulty in this approach is that the position of the centre of gravity changes with the laden condition, altering all the calculations and forcing the designer into yet another compromise.

By now, it should be clear to the layman (and familiar to the expert) that nothing in a suspension system can be considered in isolation. As well as the primary behaviour in bounce and pitch, a sprung vehicle is subjected to centrifugal forces in corners which act laterally on the mass centre to generate a roll moment about the roll axis, determined by the suspension geometry at each end. Add bounce and pitch to roll and some unexpected diagonal motion is often generated if the designer has not considered fully every combination of loadings likely to be encountered.

One theoretical study of vehicle behaviour over bumps led to the imposition of speed-reducing obstacles by road authorities in certain hazardous locations. Road humps (also known as 'sleeping policemen') were installed to excite the natural frequency of cars' suspensions and cause discomfort to the occupants. The humps in profile measured 12 ft (3.66 m) by 4 in. (10 cm), which at first glance seemed like an easy task for a modern car with supple suspension. The catch came in the shape of the hump, which was carefully developed to be a segment of a circle with a radius

When damper force is plotted against displacement, a hysteresis loop results, offset in the bump direction

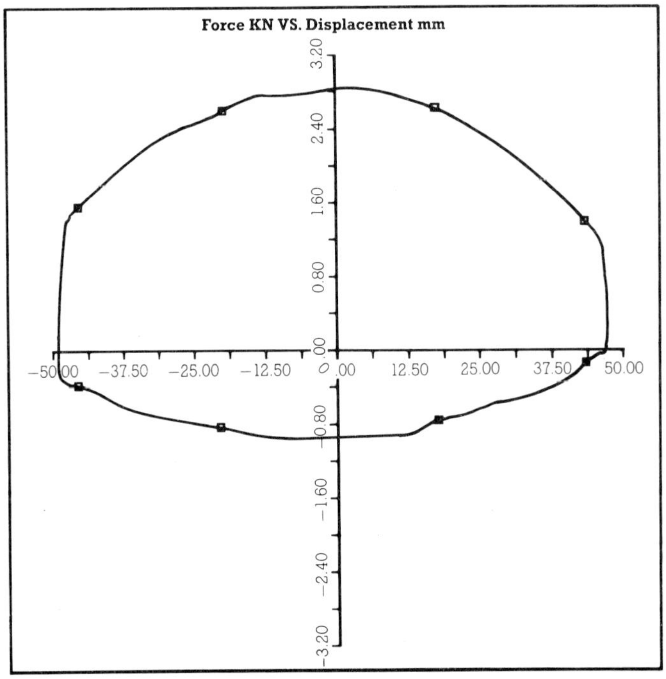

of about 18 ft (5.5 m). When encountering a hump of these dimensions and shape, a suspension with a natural frequency of between 1 and 2 hertz (the range covering most modern cars) is extremely speed critical in its reaction and hence the discomfort caused by the hump. Experiments showed that a typical car crossing a hump at anything above 20 mph (32 km/h) would generate subjective discomfort levels at least five times greater than at speeds below 20 mph.

Limiting ripple length, or the pitch between bumps that allows a suspension system to operate without reaching the limits of its travel, can be calculated easily for any particular car. The choice of 12 ft by 4 in. in the UK was based on an extensive study carried out in the early 1970s by the Transport and Road Research Laboratory (TRRL Report LR 597) in which 14 different sized humps were tested with seven different types of vehicle at speeds between 5 and 70 mph (8 and 112 km/h).

In general, it was observed that short humps gave a sharp jolt to the suspension with considerable impact loads and the risk of underbody damage as the wheels straddled the obstacle, especially with long-wheelbase vehicles. Long humps caused body displacement instead of rapid deflection of the tyres and suspension, inducing enough disturbance for the hump to be effective without the suspension crashing through to the bump stops.

4 Modern suspension systems

Understanding the theory and principles of suspension design is only the starting point in the design or analysis of any particular system. Many of the differences between systems are caused by the development of the details rather than the geometric differences in the fundamental layout. In a word, the execution of a design is worth far more than the concept.

Suspension engineers responsible for both production cars and competition machines are highly-trained, experienced specialists in a field where the amateur, no matter how enthusiastically he tackles the task, is unlikely to achieve much in the way of satisfaction should he wish to make modifications or improvements to his own vehicle. That said, the potential for the development of an advanced suspension system depends entirely upon the original design. Sometimes changes in the emphasis or balance of a standard system can be made effectively, provided the inevitable compromises are acceptable and that no vicious tricks are introduced as a result.

Initial changes to the front and rear suspensions of the Sierra by Ford Special Vehicle Engineering, for example, to increase the handling bias on the high-performance RS Cosworth version, involved replacing the front rubber bushes with solid plastic to take out compliance and improve steering response. The result was a car so quick to turn that it became 'nervous' on the road, seeming to swing its back end instantly when the wheel was moved. Before production of the 5000 examples required for Group A homologation began, modifications were made to the bush stiffness and steering ratio to reduce this effect.

There are many other aspects of interrelated characteristics where improvements in one area cause severe deterioration in another, with the inevitable compromise between ride and handling being the most familiar. Lower a car's ride height to reduce roll, and the suspension will crash through to the bump stops more readily. Replace standard springs with stiffer rates, and the ride will be firm and harsh. Fit wider wheels and tyres, and there will be heavier steering loads, more kickback on bumps and increased strain on wheel bearings and hubs.

All these conflicting variables weigh heavily in the minds of most suspension engineers when they design and develop original systems. In a way, their task is much tougher than that of an after-market specialist because they have a greater freedom of choice, in the initial design stages at least. As a result, suspension systems have shown some very diverse concepts over the years, using several different springing media and a variety of different geometrical links. Lately, there has been a more consistent gravitation towards MacPherson front struts and independent, semi-trailing rear links.

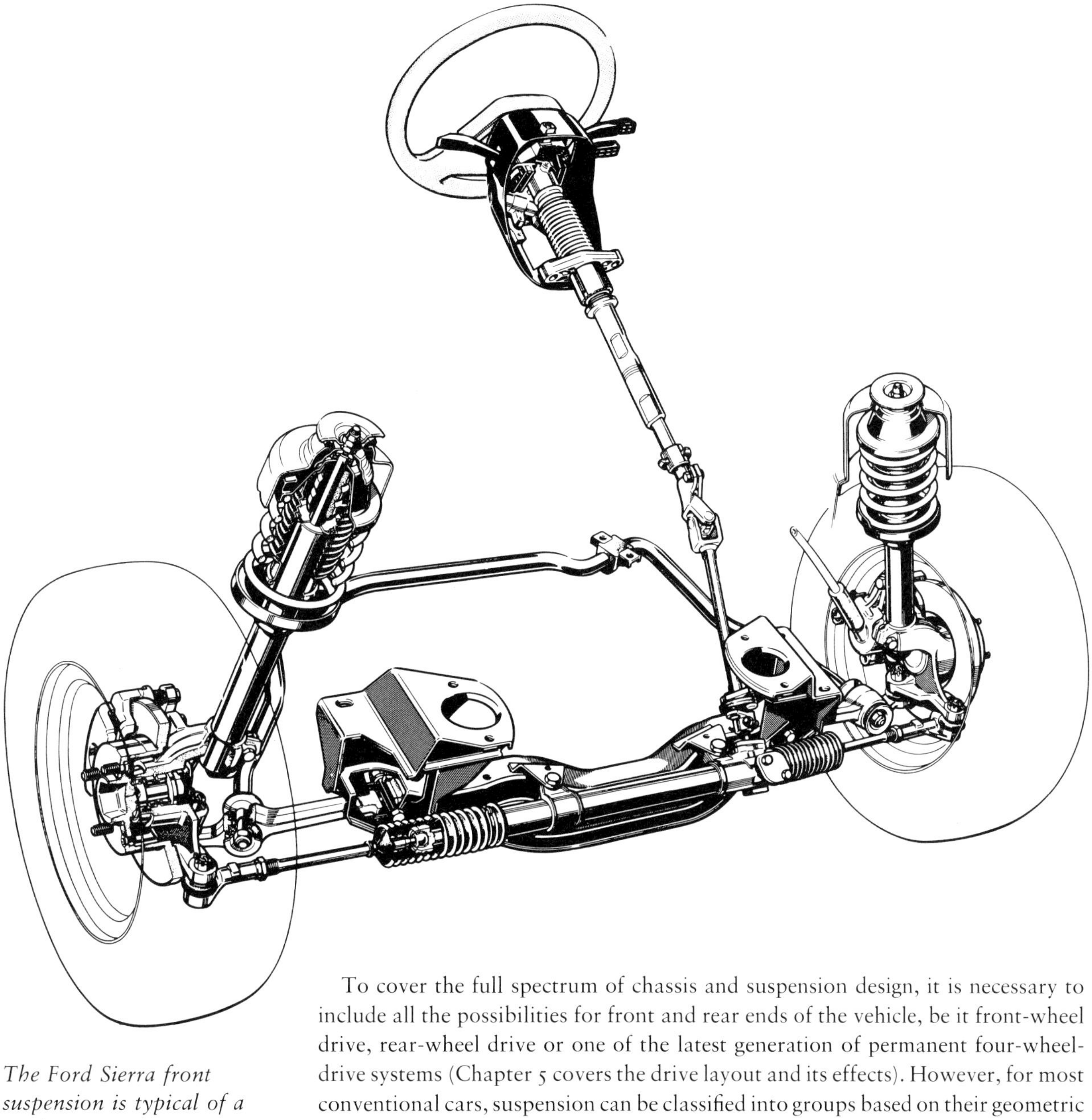

The Ford Sierra front suspension is typical of a modern MacPherson strut system which uses a lower track control arm braced by a stout anti-roll bar

To cover the full spectrum of chassis and suspension design, it is necessary to include all the possibilities for front and rear ends of the vehicle, be it front-wheel drive, rear-wheel drive or one of the latest generation of permanent four-wheel-drive systems (Chapter 5 covers the drive layout and its effects). However, for most conventional cars, suspension can be classified into groups based on their geometric elements, which are explained below.

Springing media

At the front, most cars use coil springs with either double wishbones or MacPherson struts to locate the wheels and control the geometry of their movement over bumps and in corners. Leaf springs (both front and rear) have become almost extinct due to their high weight, poor location and inter-leaf friction. The Morgan and the four-wheel-drive Fiat Panda are among the few cars still in production that use leaf springs.

Unequal-length wishbones or slightly-inclined struts are preferred now because, apart from separating the effects of the springs and dampers more effectively, they

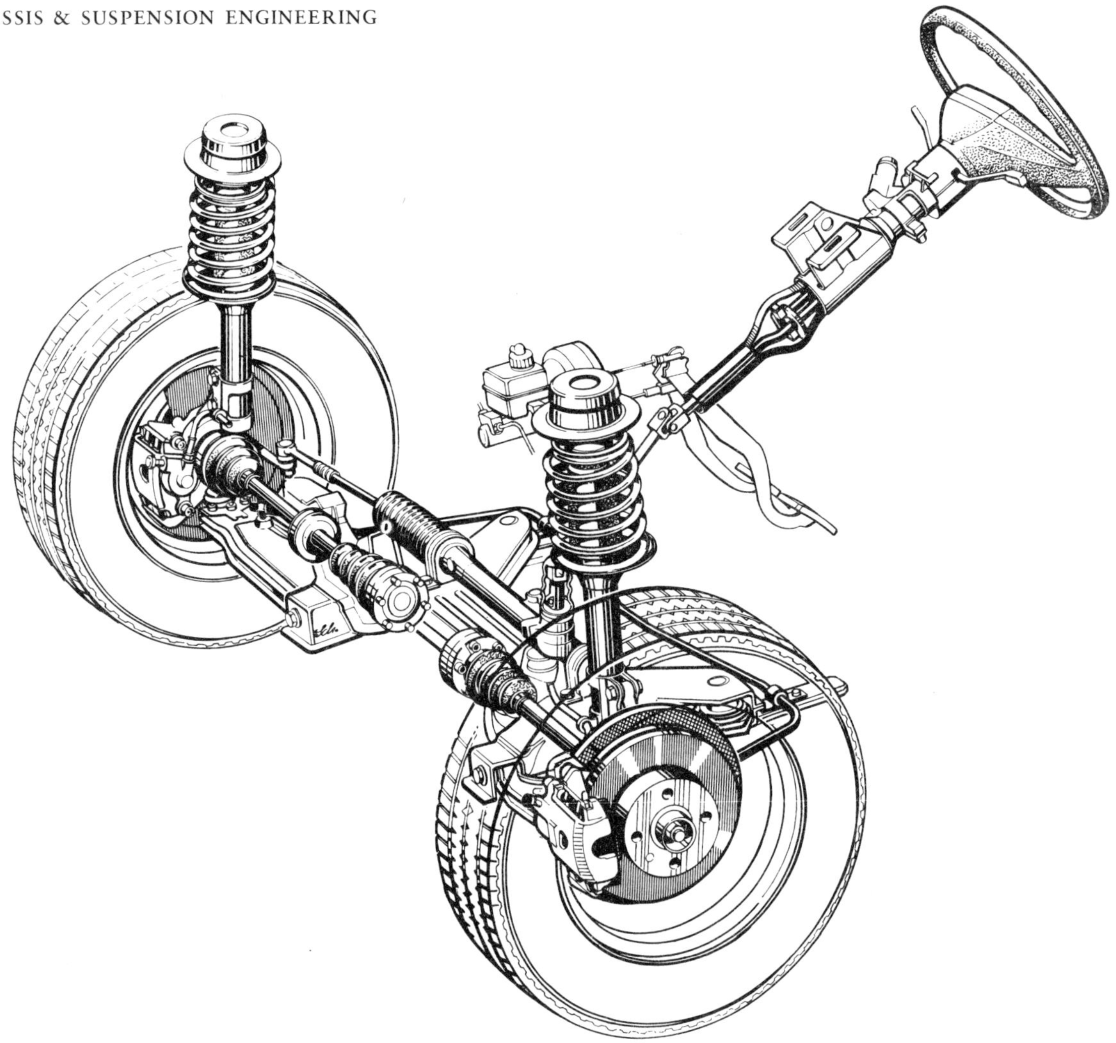

With front-wheel drive, as on the Volkswagen Golf system shown here, the lower strut attachments must be higher, to leave clearance for the drive-shaft joints. Otherwise it is very similar to that of the Sierra

control wheel movement better. Ever since camber-angle changes began to play such an important part in handling characteristics, systems that could be designed to use camber to the best advantage took over. Wheels which move up and down through a vertical path (such as those controlled by sliding pillars, equal-length wishbones or pure leading or trailing links) take up the same degree of positive camber in corners as the body roll, reducing cornering power in the process.

Struts are usually preferred to wishbones because they are more efficient at spreading the load inputs to the body, and in most cases eliminate the need for a heavy subframe to provide acceptable insulation from noise, vibration and harshness. Wishbones at the front of rear-wheel-drive cars are usually more compact, allowing a lower bonnet line, but when drive shafts are added for front-wheel drive, the lower spring attachment moves up above the hub to avoid component conflicts.

MacPherson struts

The most commonly used design of front suspension on modern cars is known as the MacPherson strut. It is named after an American Ford suspension engineer called Earle S. MacPherson, who, nearly 40 years ago, patented the idea of locating the lower end of an inclined strut system by means of the anti-roll bar. Although,

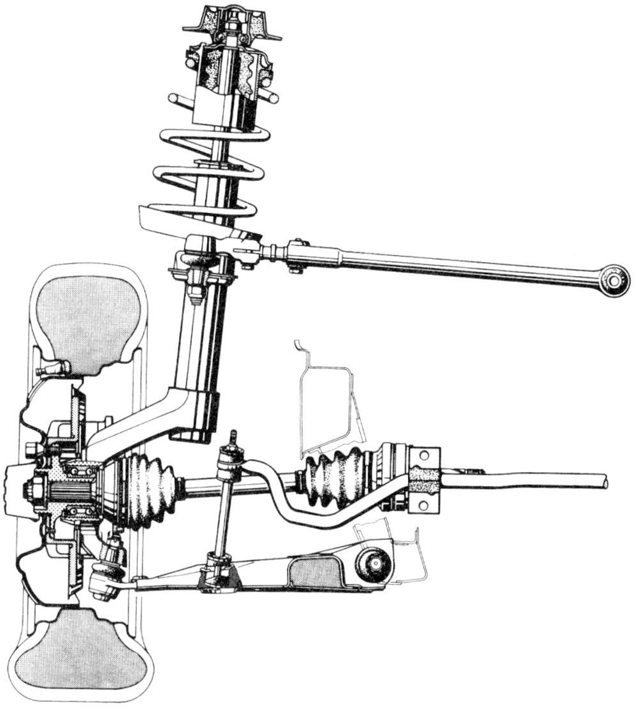

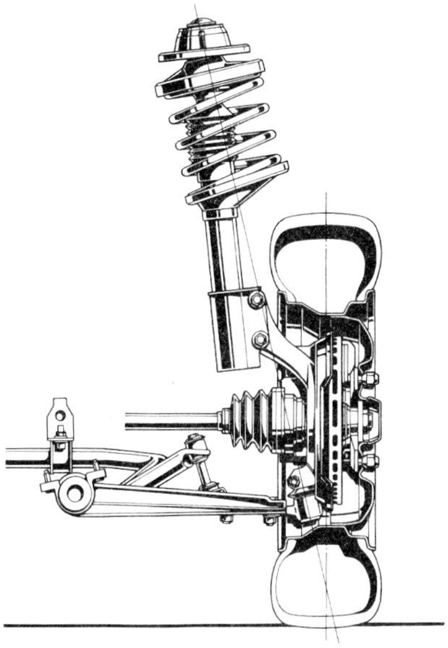

strictly speaking, a description of only this composite system, the term is applied to all kinds of strut suspensions.

A MacPherson strut is based on an outer tube which is fixed to the hub carrier at its lower end and welded to a seating cup for the suspension spring at its upper end. Inside the tube is a telescopic damper, the piston rod of which is attached to a thrust bearing in the centre of a turret formed in the car's inner wheel-arch area. This also acts as the upper spring seat. Location of the lower end of the strut is triangulated, usually by a transverse link (or track control arm) and a tie bar or the lever arm of a transverse anti-roll (or stabilizer) bar.

The differences between the front suspension systems used on the Vauxhall Astra (left) and the Mazda 323 (right) are only slight. In both cases, the bottom of the strut is located by a wishbone and the anti-roll bar is connected by a short vertical link

The advantages of a MacPherson strut are that the suspension inputs to the body can be directed to inherently stiff and well-spaced areas of the structure. Like the alternative unequal-length-wishbone systems, struts allow considerable design flexibility, but with the significant advantage that a front subframe is not usually required to ensure adequate close tolerance for the location points.

Double wishbones

There are several ways in which pairs of parallel arms can be arranged to control suspension geometry. There may be an upper and lower trailing arm on each side, used with transverse torsion bars as on the now classic VW Beetle layout,

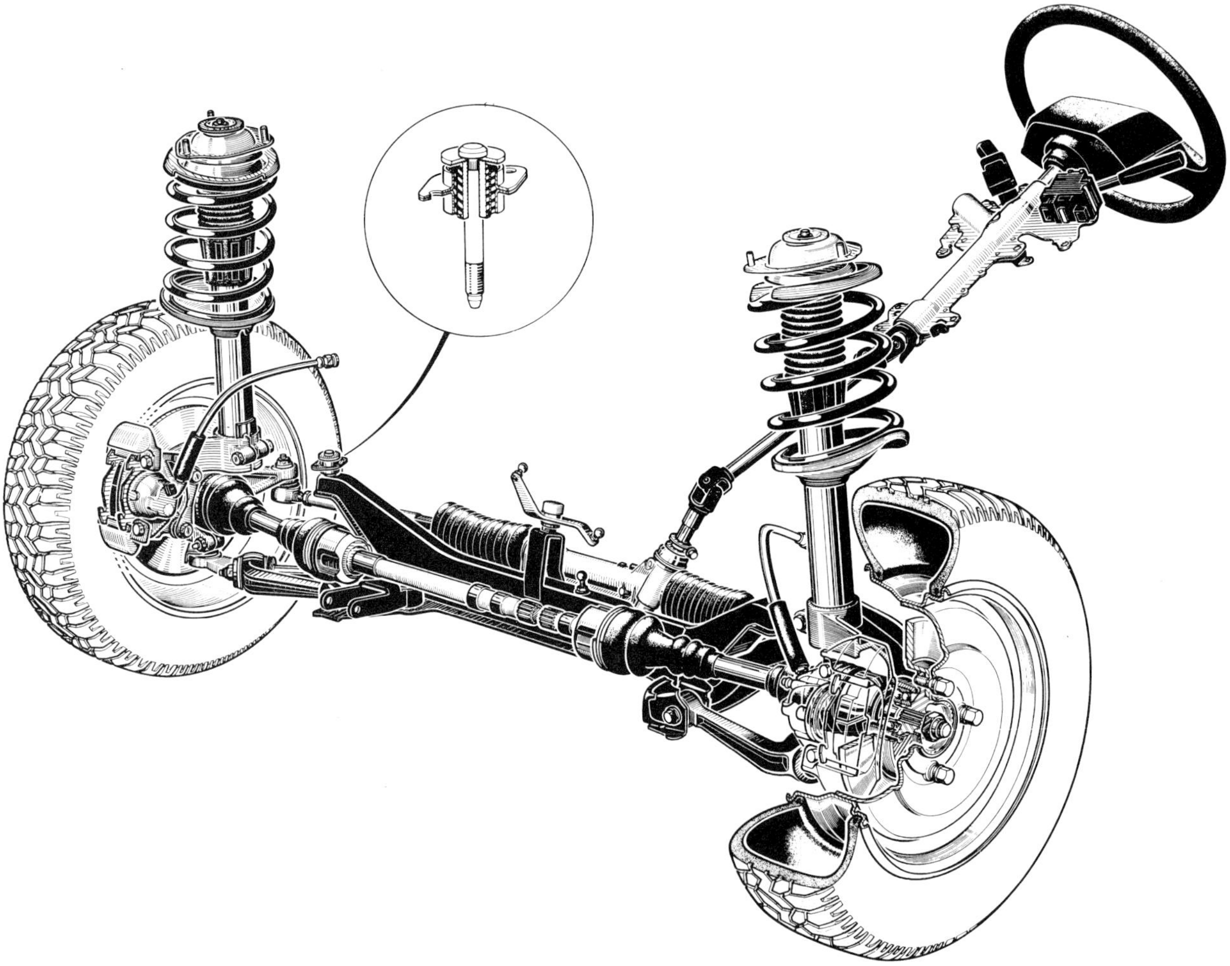

ABOVE *In the Peugeot design (shown here for the 309 model) there are dog-legs in the lower arms to control the compliance of the anti-roll bar bushes more precisely*

RIGHT *Almost the last double-wishbone front-suspension system left in production, the design of the new Jaguar has been developed from previous layouts. It is particularly compact and extremely efficient in its execution*

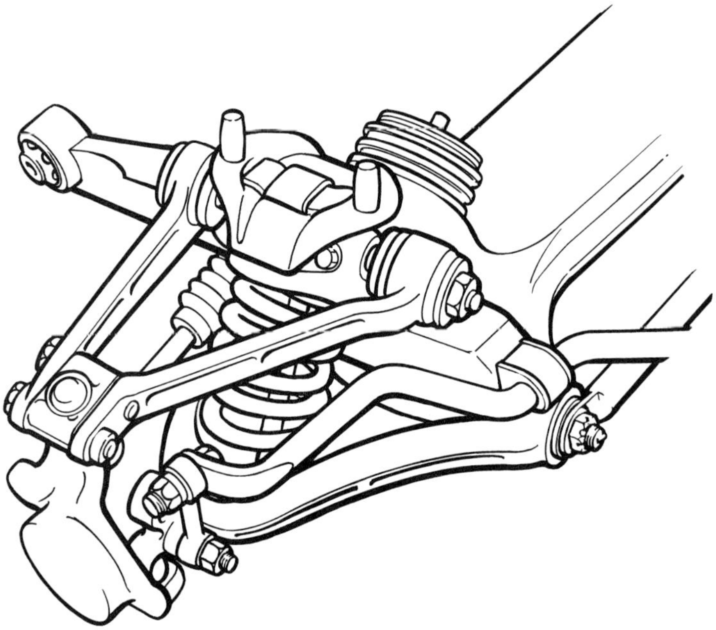

symmetrically-arranged, transverse A-frame links or angled derivatives and hybrids. Parallel motion from equal-length links maintains wheel camber angles at their static value over bumps, but causes track changes and positive camber angles to be generated by body roll, to the detriment of cornering ability.

A better way to control camber is to make the upper links shorter than the lower ones, so that the top of the wheel moves in as the suspension is displaced from its normal travel position. In many applications, when lever-arm hydraulic dampers were popular, the short upper link was formed by the damper-actuating lever.

If a substantial front subframe forms part of the vehicle design, double wishbones can be designed for better attenuation of road noise. Close manufacturing tolerances are vital for precision control of the system, and it is usual to machine the attachments after subframe assembly. On the Jaguar XJ6, which is a supreme example of how well a wishbone system can be developed, the dampers are attached directly to the body structure to ensure that the damping forces on the body are not reduced by the insulation of the subframe mounts.

Independent rear-suspension systems

The considerations governing the choice of rear-suspension design are highly dependent on the drive configuration (see Chapter 5). The choice for the rear end of a car is simple: there can be either a beam axle with a wheel attached to a fixed hub at each end or an independent arrangement. Just as the steering system and drive arrangements for front-wheel-drive layouts add complexity to independent wheel movement at the front, the drive system alone is the complicating factor at the rear (except in the case of rear-wheel steering systems, see Chapter 11).

In general terms, the case for independent rear systems with rear-wheel drive is easy to prove, whereas its justification with front-wheel drive is not. A live rear axle system, using a conventional rear axle, coil springs and proper location by double links each side and a transverse Panhard rod, typically provides about twice the

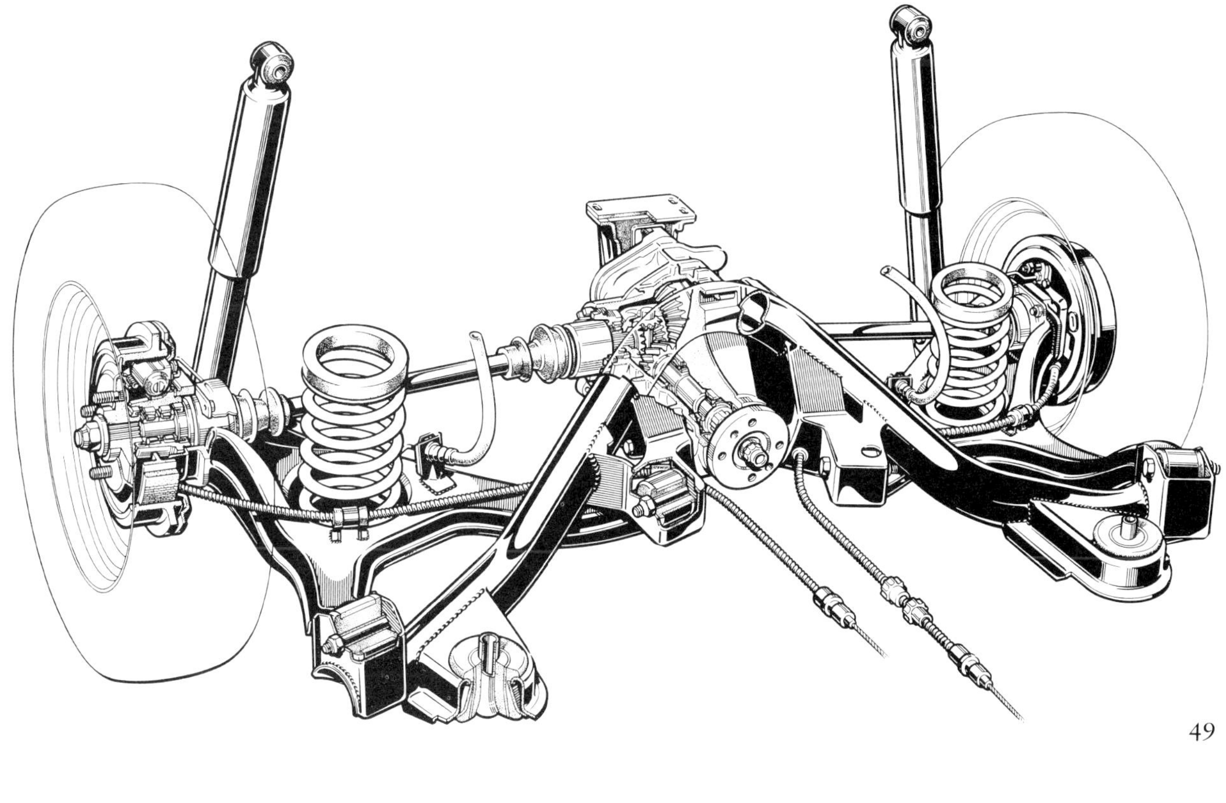

The Ford semi-trailing-arm irs used on the Sierra is attached via a tubular subframe to improve road noise insulation

unsprung mass of an independent arrangement. In between is the undriven, dead beam axle and the semi-independent de Dion system (a de Dion axle has the differential and final drive mounted on the chassis with the two wheels connected rigidly by a lightweight, transverse tube, often curved to clear the axle housing and hollow to save weight and provide drive-shaft access to the hubs).

Variety in driven rear-suspension design used to be more diverse, with many different live-axle layouts and fully independent systems located by many different kinds of geometric linkage. The simplest, but notoriously tricky type of independent system is the swing axle. This type was used by Volkswagen on the original Beetle and all of its later derivatives until the last Super Beetle version in 1972 when a semi-trailing arrangement took over. Renault employed it on the 750 4CV, Dauphine and all their later rear-engined designs, Chevrolet on the rear-engined Corvair, Porsche on their rear-engined dynasty, Mercedes-Benz on the classic 300SL Gullwing, Triumph on the Herald, and many more besides. Swing axles were particularly favoured with rear engines in cheaper family cars because they eliminated the need for costly, outboard universal joints.

The problem with the swing axle was that its roll centre was relatively high, causing the cornering forces at ground level to twist the system, lift the rear end and flick the wheels into positive camber, reducing their grip dramatically. With the tyre sections available at the time, swing axles usually displayed very unpredictable cornering behaviour.

Mercedes-Benz started a significant trend when they turned away from the swing-axle concept in 1972. Prior to the S-class launch that year, Mercedes-Benz had stuck resolutely to various designs of swing-axle rear suspension. They liked the way swing axles helped large, front-heavy cars negotiate tight curves. It was a heritage derived from the early days of racing, when the fastest way through a turn was a four-wheel drift.

Over the years, they moved away from the swing-axle characteristics, first by lowering the pivot point to reduce the roll moments and hence the camber changes, then by changing to a totally different suspension geometry in 1972. By the time they reached this stage of development, they had tamed the swing axle to the point where it functioned extremely well under normal ride conditions. But it always had a vicious final sting, literally in its tail, when the unexpected happened and the driver was forced to brake in the middle of a bend.

Partly to handle the extra power of their largest-engined cars and partly to avoid any risk of the handling-related, product-liability issues that Ralph Nader was drumming up in the USA around another swing-axle car (the Chevrolet Corvair), the S-class rear end became a more or less conventional, semi-trailing-arm system. For road and axle noise insulation, it was mounted on a V-shaped, large-section, fabricated-steel subframe which was supported on three large-diameter, rubber bushes. Widely-spaced pivots for the angled rear wishbones had little compliance, but the upper spring and damper attachments were made through thick rubber insulators. Half shafts were fully articulated with constant-velocity, plunging joints at each end, and there was a two-piece propeller shaft with centre steady bearing. Linear-rate springs were near vertical, concentric with the telescopic dampers, and there was a cranked anti-roll bar mounted above the wheel centreline.

Careful arrangement of the rear geometry introduced an anti-squat feature, to resist weight transfer under hard acceleration, while at the front there was an equally-clever anti-dive system, also described as anti-lift when mentioned in relation to the rear end. In both systems, the torque at the wheels was arranged to

Jaguar's irs for the new XJ6 range is built around the final-drive unit and based on a triangulated subframe with pendulum links

generate vertical suspension control by angular displacements of the suspension pivots, countering the effects of dynamic mass transfer. The harder the driver accelerated, the more the effect, giving a very reassuring feel under full power or emergency stops.

However, it is the packaging advantages of semi-trailing-arm systems as much as their dynamic control of wheel geometry which has made them a firm favourite since, with only a few specialists like Jaguar sticking to their traditional, double-wishbone designs.

The fundamental deficiencies of semi-trailing geometry are mostly limited to the changes that occur in camber and toe setting with wheel travel. The need to make the wishbone bushes soft enough for acceptable primary compliance also causes some secondary problems of throttle steer (toe change caused by power-on/power-off settings). Reaching the optimum for the disturbing effects of squat and lift is a more difficult task.

On the other hand, double independent wishbones provide total flexibility of static geometry and geometry changes with wheel travel. They offer the possibility of full squat and lift compensation, and their unsprung mass is relatively low, especially if component economies are made (as on the Jaguar where the drive shafts provide wheel location).

In the first design of Jaguar rear suspension, a forward-facing radius arm was used on each side to assist in control of pitch and yaw of the unsprung masses and the subframe assembly. The wishbone pivot points and the positioning of the radius arms allowed little possibility of controlling the squat and lift characteristics. In the development of the later XJ6 system, the pivots were inclined,

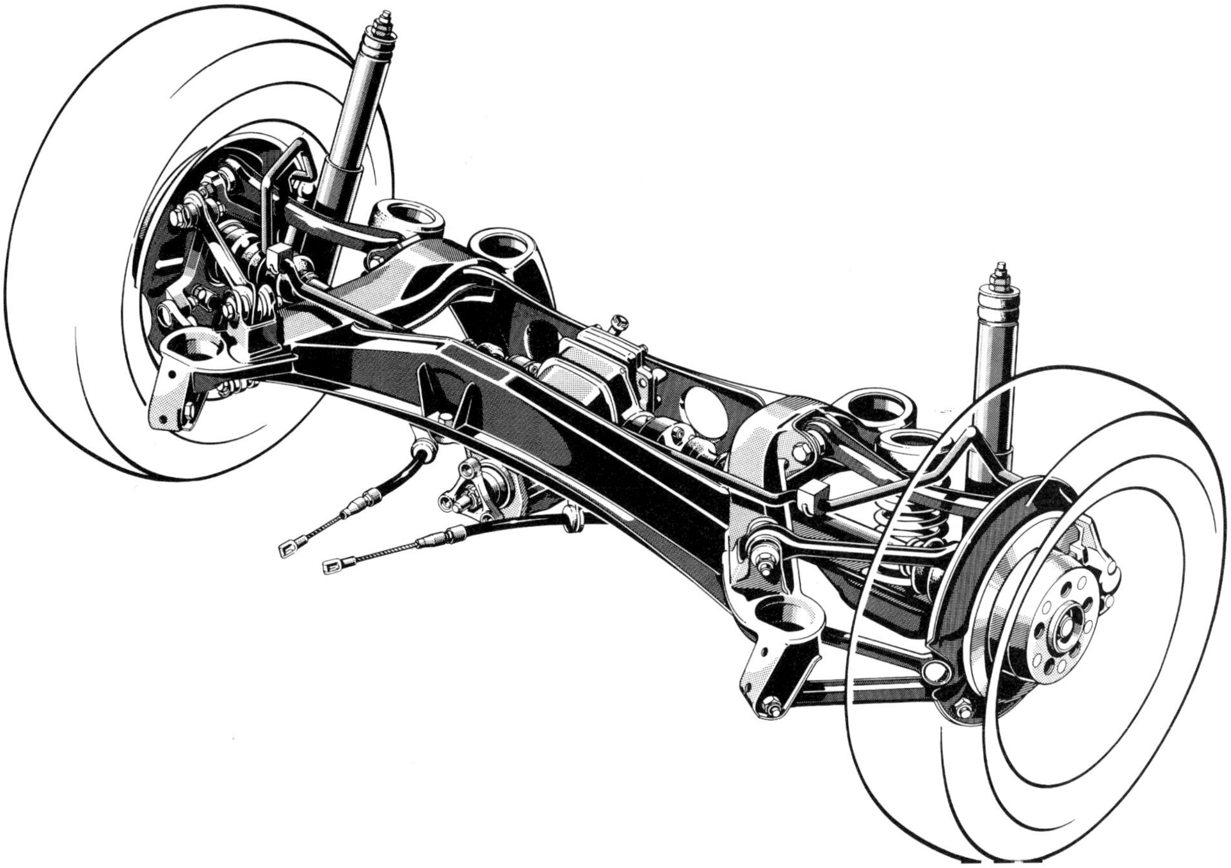

The multi-link rear suspension for the Mercedes-Benz 190 range incorporates anti-squat and anti-dive geometry. It takes up more space than conventional semi-trailing systems, compromising rear seating space

but the resultant path locus of the hub with wheel travel was incompatible with a simple, horizontal radius arm. Substituting a Watts linkage, which could handle the geometry, was ruled out on grounds of cost and complexity.

Several other advanced rear systems were also tried before the final solution was selected. It is a double-wishbone system much like the previous one, but the inner pivots for the lower arms are mounted on what is described as a pendulum, or hanger yoke, at the front of the differential casing and a straight cross-tie at the rear. It provides inboard compliance that is independent laterally from the longitudinal stiffness, so that bump shocks can be absorbed independently from the transverse control of the suspension arms.

Semi-trailing-arm, independent rear suspension does not provide as much design flexibility in selecting the amount of wheel camber change and toe change that occurs during roll in corners, and hence the amount by which the rear cornering power is increased. For each trail angle chosen, there will be a specific amount of toe change, the effects of which must be considered in association with the length of arm required, the inclination of the pivots to the horizontal plane and the various influences of compliance in the bushes. Trail angles of around 23 degrees on the Ford Granada were reduced to 18 degrees on the Ford Sierra to give optimum handling in terms of power-on/power-off response, steady-state cornering and transient behaviour.

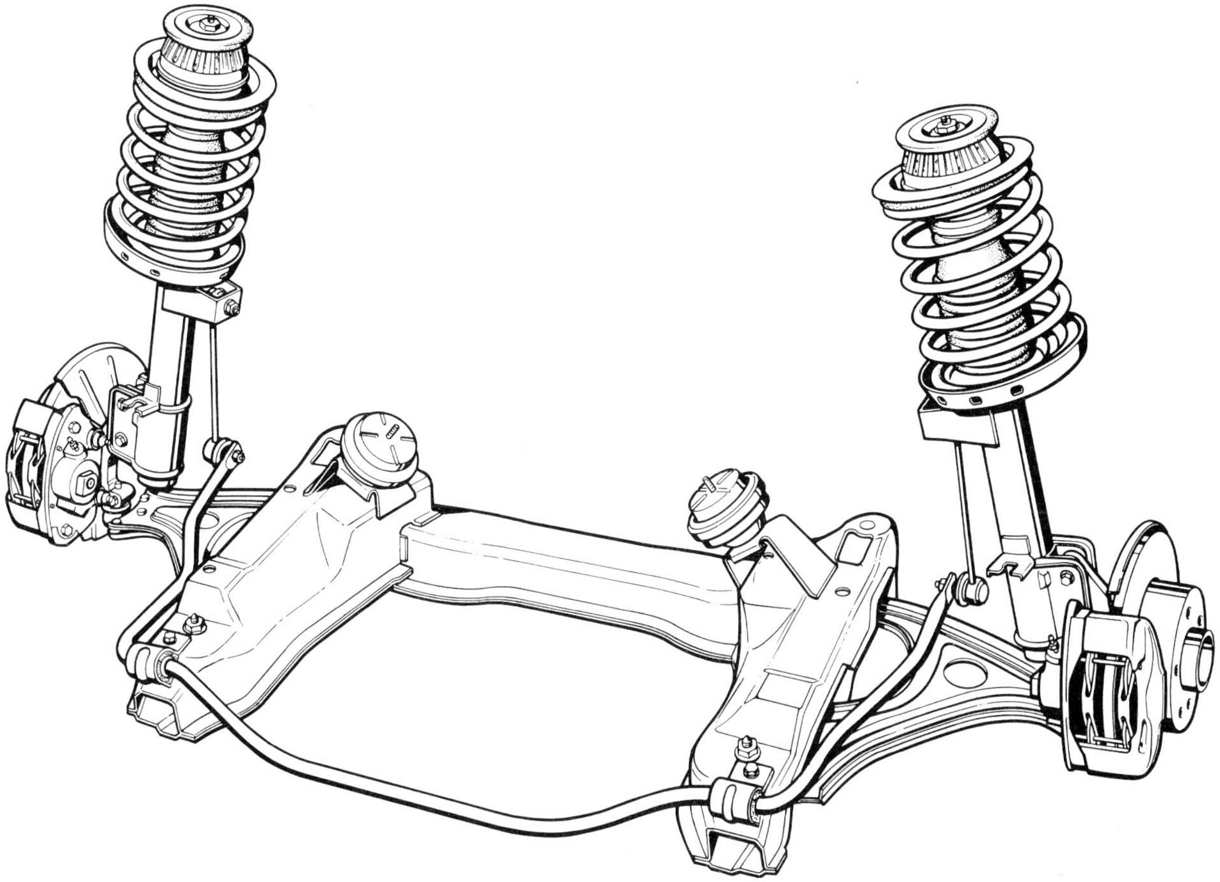

Front suspension for the new Opel Omega (Vauxhall Carlton in the UK) is designed with negative offset and a self-stabilizing steer system functioning through the inner wishbone bushes

Opel's self-stabilizing system

Suspension engineers working on the Opel Omega (Vauxhall Carlton in the UK) devised some extremely advanced characteristics for the fully independent system that cause it to react automatically to stabilize the handling and improve safety levels in severe conditions. In many ways, it comes close to being an 'intelligent' suspension that thinks for itself and initiates its own corrections.

At the front, the basic element of the concept is a strut-and-wishbone system incorporating a unique combination of negative-scrub-radius geometry and compliance steer that gives a self-stabilizing effect under uneven braking forces. When a front wheel tries to turn outwards, the geometry of the insulating bushes forces it back to the true running angle.

At the rear is a fully independent, semi-trailing-arm system with a compound trail angle (inclined in two planes) designed to improve handling and cornering power, operate more consistently over the full range of load conditions, reduce tyre wear and introduce a small amount of rear-wheel steering when required to improve stability in extreme conditions.

Mechanically, the front suspension takes the form of a conventional strut system, angled inwards at the top and located at the lower end by a fabricated, A-shaped, steel wishbone. The damper has a wet piston with a larger effective area and better heat dissipation than the alternative dry cartridge type, to provide improved damping under extreme conditions. It is clamped at its lower end to a forged, one-piece steering knuckle and hub employing sealed, maintenance-free wheel bearings

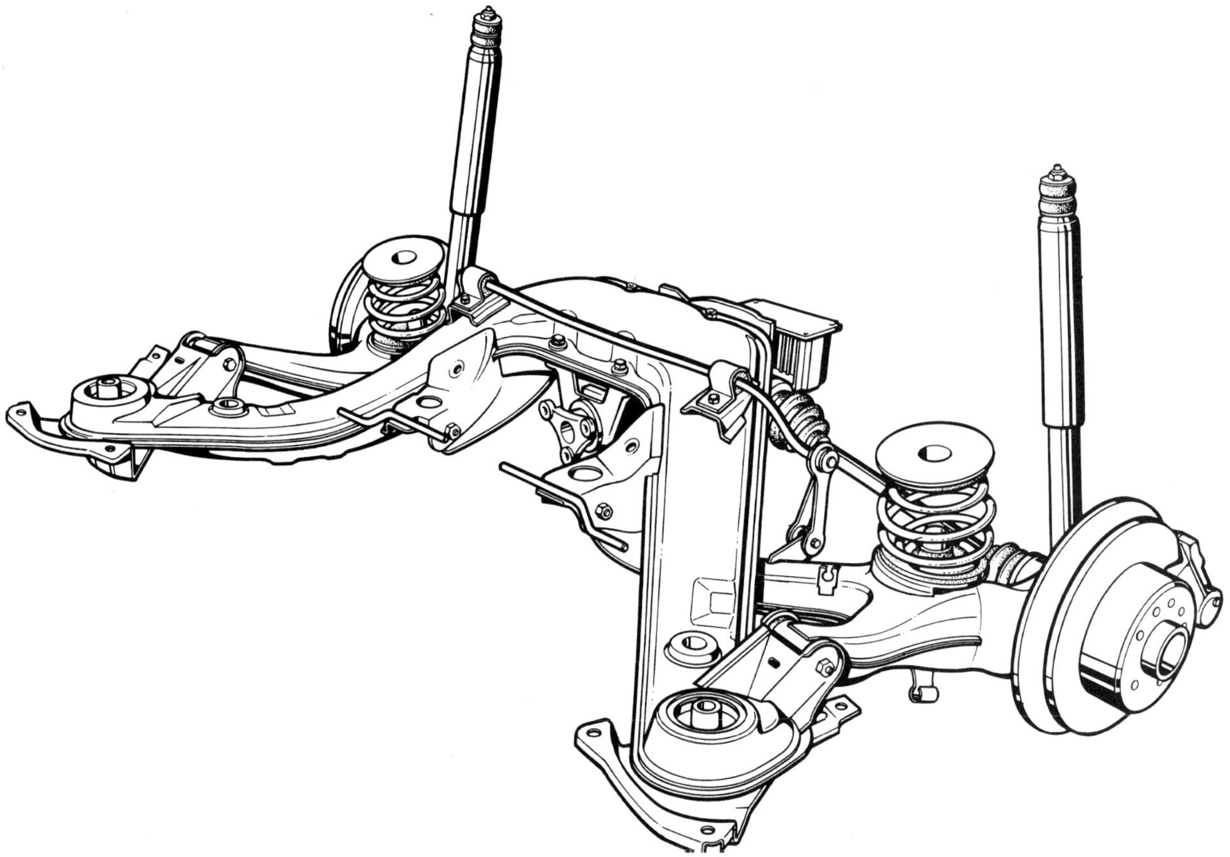

At the rear of the new Opel, compound trail angles are used to improve vehicle stability through induced toe steer

and a small range of camber adjustment (± 20 minutes, or $\frac{1}{3}$ degree) to ensure that every car has the same precise settings in production.

At its upper end, the strut has a new 'decoupled' mounting system that insulates the body structure more effectively from road shocks by separating the load paths of the combined concentric spring and damper unit. At the lower end, the centre of the tyre contact patch is arranged to be $\frac{1}{25}$in. (1 mm) inboard of the point where the extended steering axis intersects the ground. This provides a negative scrub radius that tends to counter any imbalance in the braking or frictional forces at the front wheels.

Directional instability when braking on surfaces where the coefficient of friction is not equal on both sides of the car (as with one wheel on ice and one on wet tarmac) is usually caused by the natural tendency for the wheel with the better grip to deflect its suspension bushes more and turn outwards slightly. The Omega counters this by a novel wishbone-pivot arrangement: the rear bush is installed horizontally and the front one vertically. Therefore, under the forces generated by the negative offset, the wheel with the better grip turns inwards instead of out, compensating for the extra drag on that side of the car and helping to maintain a true course. Because the degree of self-stabilizing toe-in is proportional to the magnitude of the disturbing force, the system operates automatically with its own level of artificial intelligence.

The rear suspension of the Omega is also designed for maximum stability, having independent kinematic control to provide high cornering power combined with automatic compensation for the normal unsettling changes that occur during transient manoeuvres. For maximum grip in a turn, a tyre should lean in at the top to give negative camber. The problem with a semi-trailing, independent system is that the camber angle changes with suspension movement under both dynamic roll forces and the static laden condition, affecting the handling and increasing tyre wear. The need for softer suspension bushes at the rear to improve ride comfort and reduce the harshness of modern, low-profile tyres also introduces compliance-steer effects induced by the forces at the wheel when decelerating or braking. These conflicting effects were reduced on the Omega by arranging the rear suspension arms to pivot at a compound angle (displaced from the longitudinal axis of the car in both the horizontal and vertical directions). The effect is to increase toe steer when lifting off to maintain a constant rear roll height (the point about which the body rolls) under all conditions.

Opel pioneered a new trend towards lower rear trail angles with the independent suspension system introduced on the Senator in 1977. The Omega system is a new-generation development of the same concept, the trail angle in the horizontal plane being reduced still further from 14 to 10 degrees and, for the first time, combined with a 1.35-degree vertical inclination. This allows consistently higher negative-camber angles to be maintained throughout the full range of suspension travel for improved stability and handling without increased tyre wear.

While most conventional trailing-arm systems can be designed to perform adequately under typical two-passenger load conditions, they are restricted in practice by the need to stay within the tyres' durability limits at the maximum rated, gross vehicle mass. The height of the rear roll centre varies with the load as well, tilting the roll axis (the line joining front and rear roll centres about which the whole car rolls) from a stable downward inclination to a less stable upward inclination when fully laden.

On the Omega, rear camber angles and the rear roll height remain virtually constant at all times, while the front roll height reduces slightly with load, increasing stability under conditions when most conventional cars start to feel unsteady. In addition, the vertical trail angles induce a small degree of toe-in under deceleration and braking, countering the normal tendency for the tail to swing out when lifting off in a corner. These combined characteristics provide a measure of 'response steer' that automatically stabilizes the Omega at the rear with the same kind of artificial intelligence as that applied to the front.

Other benefits derived from the new design include reduced unsprung masses for better wheel control, multi-axis front cross-member support brackets and double-isolated anti-roll-bar attachments for improved noise suppression, minibloc progressive-rate coil spring design that prevents metal-to-metal contact at full bump, and optimized damper mountings positioned where they operate at the best leverage ratio.

For the much less demanding task of controlling the rear wheels on a front-wheel-drive car, many different systems still exist, ranging from the simple beam axle and struts of the Ford Fiesta to the elaborate rear wishbones of the Renault 25. Transverse links, located longitudinally by tie bars (as used on the Escort) might offer some ride advantages over the compound-beam system adopted by almost every other fwd saloon (pioneered by Volkswagen on the Golf and widely copied ever since), but tyre wear is severe as the contact patches shuffle in and out over bumps, and the overall handling leaves plenty to be desired.

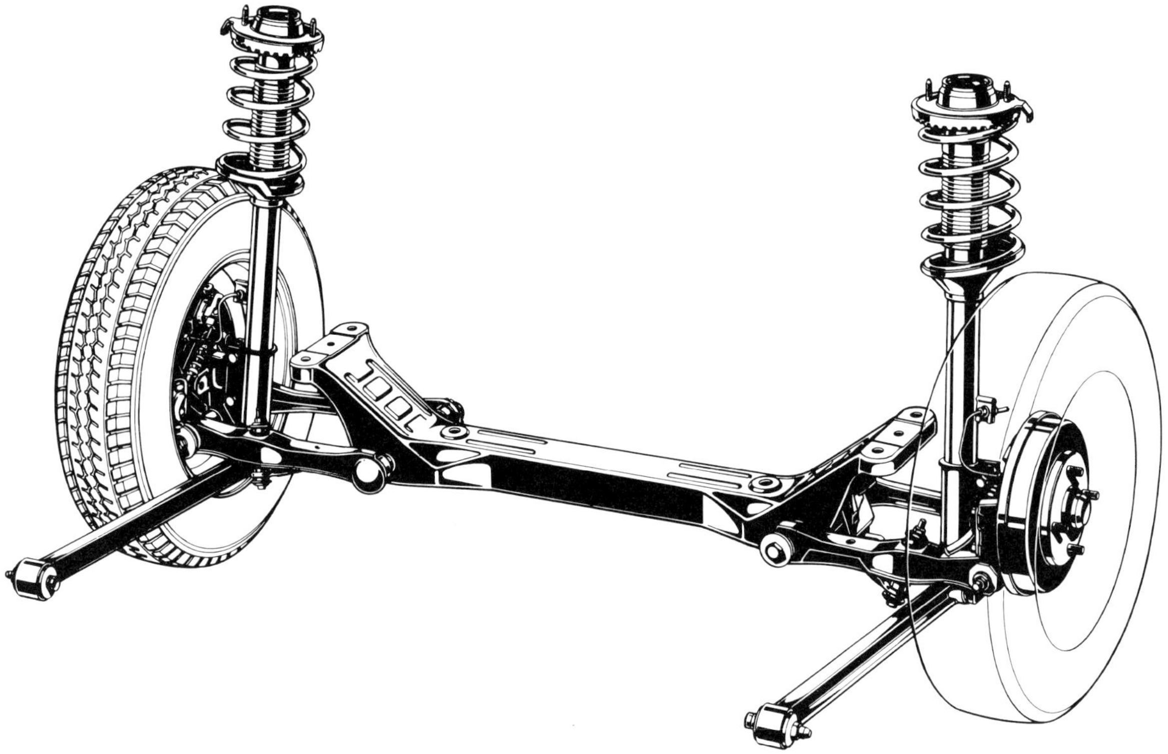

Like the Ford Escort, the Mazda 323 rear suspension is unusual for a small front-wheel-drive car in being a truly independent system. It uses double transverse links each side, braced by long trailing arms

ABOVE RIGHT *The VW Golf was one of the first cars to perfect the trailing twist axle, a design that has become increasingly popular on small front-drive designs. The open-sided cross-tube is designed to resist roll forces by twisting*

RIGHT *Peugeot's very compact torsion-bar design used on the 205 and 309 allows the dampers to be laid down at about 30 degrees, leaving more space above for luggage in the boot*

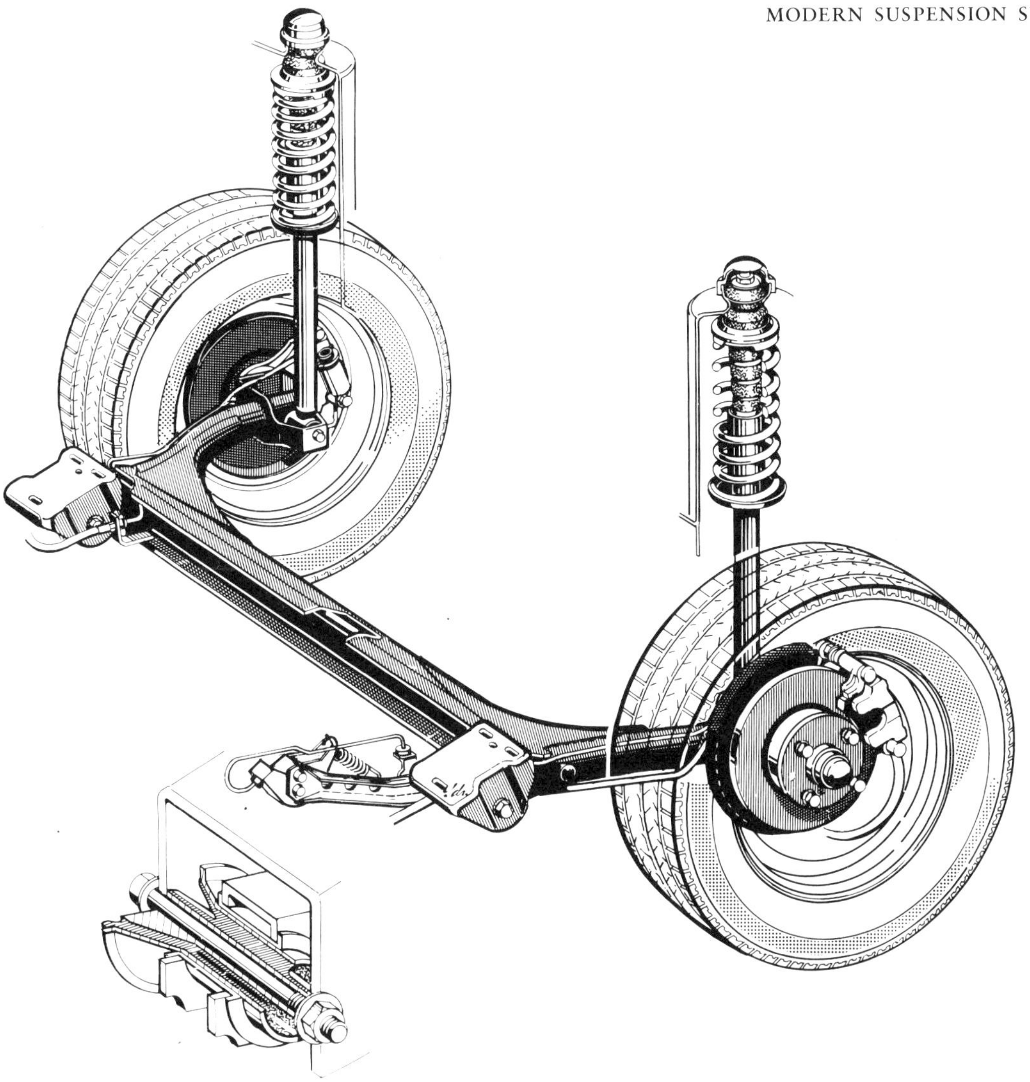

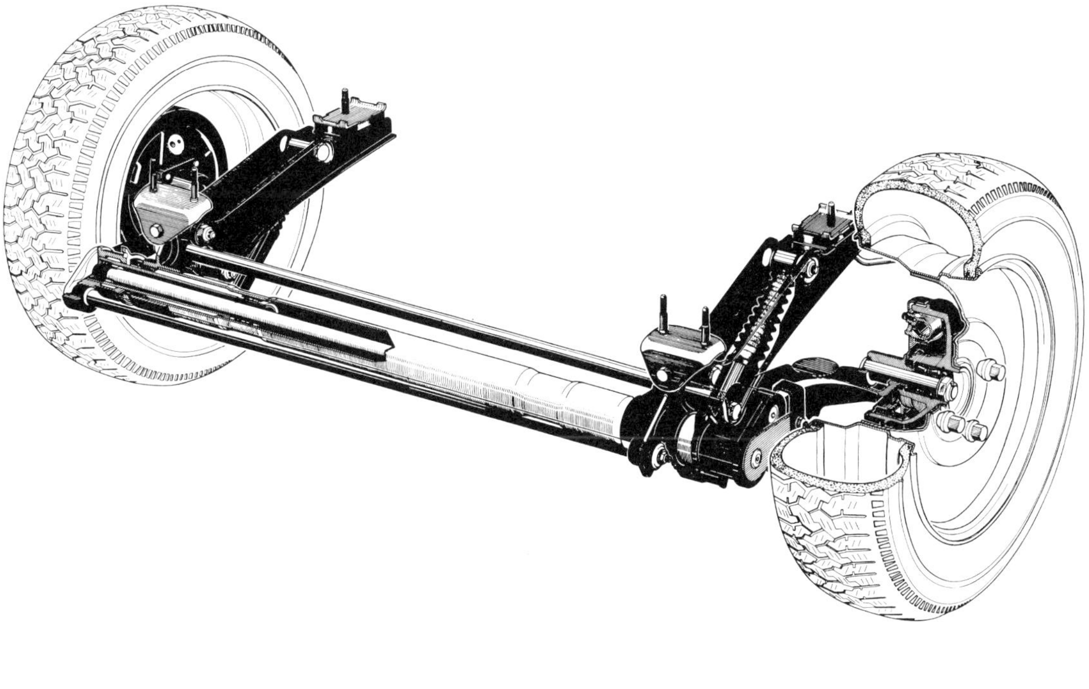

The VW system is a clever piece of technical design and uses a compound beam that is stiff in compression to provide good lateral wheel location, but is relatively flexible in torsion to induce camber changes in corners. It acts as a huge anti-roll bar, operated by the trailing arms it locates, and is extremely simple, cheap to manufacture and very effective.

Design of the compound beam, or trailing twist axle as it is also called, is a typical example of how the control of engineering characteristics is developing as a new science with significant benefits in efficiency. Undoubtedly, it will be the subject of much further development in the future.

A compound beam consists of two trailing arms interconnected by a rigid cross-tube. It depends on large elastic deformations of the structure to perform the essential roll functions for single wheel deflections, but remains unstressed when traversing a uniform ridge. There are no flexible links in the axle assembly, although some applications have a Panhard rod to control lateral axle movement.

One attraction of the compound beam, from the designer's viewpoint, is that the cross-tube location provides a versatile method of fine tuning the handling. Placed at the trailing-arm pivots, the cross-tube acts only as a conventional torsion anti-roll bar, but moving it rearwards to the wheel centres subjects it to a variety of conical and axial displacements as well, adding bending loads to the torsional strain.

The optimization of a compound-beam design, where the trailing arms must be laterally and torsionally elastic as well, cannot be achieved by manual techniques and depends heavily upon finite element stress analysis (see Chapter 6) of the complete welded assembly. The selection of the geometry for bump steer, roll steer, roll camber, deflection steer and the roll-centre height also requires the use of advanced, dynamic computer modelling.

RIGHT *In this Monroe twin-tube damper, the downward movement of the central rod forces the piston to drive fluid through the lower compression valve into the outer chamber. On rebound, the piston moves up and draws the fluid back. At the same time, the rebound valves in the piston open to compensate*

FAR RIGHT *In this Bilstein monotube, or gas-filled damper cartridge, there is a supplementary chamber containing a pressurized gas which compensates for the difference between the compression piston area and the rebound area caused by the area of the operating rod*

Dampers and how they work

As described in Chapter 3, dampers are hydraulic devices used to restrict the number of cyclic oscillations caused by a bump deflection. They do this by pumping fluid through orifices to convert the energy to heat which can be dissipated in the atmosphere. The objective in damper-valve design is to maintain consistent laminar-flow characteristics throughout the operating range of loads and input velocities. That usually means incorporating several small nozzles or jets rather than a single large one, arranged to open in stages according to the stroke and internal pressure generated to create a specifically-tailored resistance curve.

The most commonly used form of hydraulic damper is the telescopic twin-tube type which combines the action of fluid passing through a piston in a cylinder with the displacement of fluid into a concentric chamber. Conveniently, it separates the bump and rebound actions of the suspension movement by the use of separate blow-off valves.

The damper itself takes the form of a piston running in a cylinder and activated by a rod. Within the piston there are the spring-loaded bump valves which open under pressure to allow the piston to move through the almost incompressible fluid. The blow-off characteristics of these valves can be controlled by the design of their profiles and the spring behaviour.

On the rebound stroke, the effective area of the piston is reduced by the cross-sectional area of the driving rod, so as well as fluid passing through the piston on bump movement, a proportion is forced through valves at the opposite end of the

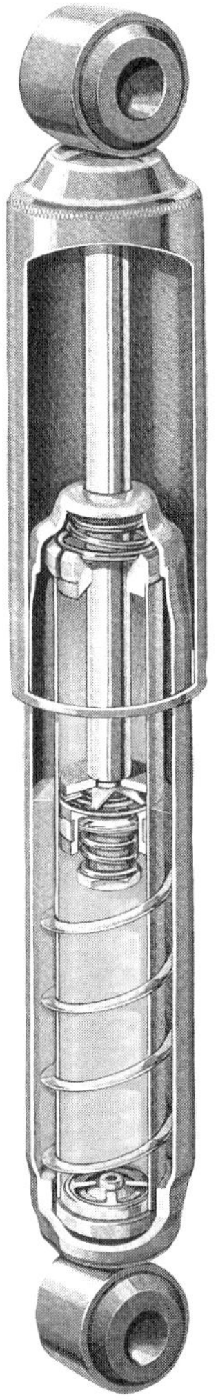

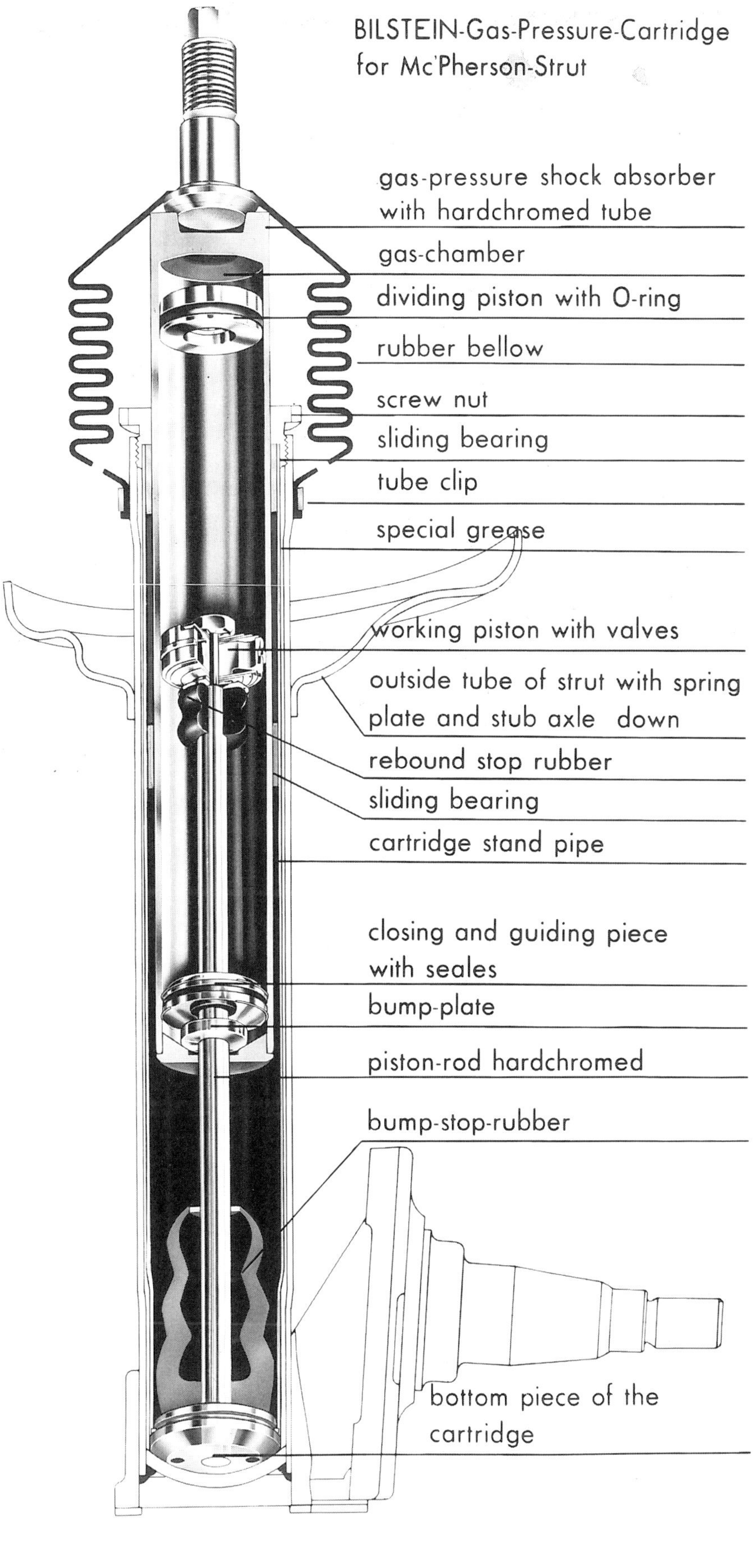
BILSTEIN-Gas-Pressure-Cartridge
for Mc'Pherson-Strut
gas-pressure shock absorber
with hardchromed tube
gas-chamber
dividing piston with O-ring
rubber bellow
screw nut
sliding bearing
tube clip
special grease
working piston with valves
outside tube of strut with spring
plate and stub axle down
rebound stop rubber
sliding bearing
cartridge stand pipe
closing and guiding piece
with seales
bump-plate
piston-rod hardchromed
bump-stop-rubber
bottom piece of the
cartridge

cylinder to compensate. Considerable volumes of fluid flow in and out of the outer chamber where they are cooled quite effectively by the air flow created by the vehicle's forward speed. The limitation of twin-tube damper performance is usually caused by aeration of the fluid during its passage to and from the outer reservoir, or by localized temperatures elevated above the fluid's vaporization point by continuous, severe bumps, creating gas pockets.

Single-tube, telescopic dampers actually predate the twin-tube variety to patents once held by de Carbon. Instead of a supplementary chamber being used to compensate for the piston rod area, a volume of inert gas under pressure is incorporated at the bottom of the cylinder and separated from the fluid by a second floating piston. For this reason, single-tube dampers are also known as gas-filled dampers.

As the main hydraulic piston of a single-tube damper is forced through the fluid under bump movement, the gas is compressed by the increased pressure on the floating piston which moves away to increase the bump fluid volume, effectively decreasing the working area. Unlike the behaviour of the twin-tube damper, the resistance curve for the gas-filled damper increases at a constant rate instead of tailing off, giving more consistent ride characteristics. There is also better overall cooling without the outer supplementary chamber and no risk of fluid aeration.

To function satisfactorily, the damper must be inserted between the moving suspension parts and a relatively rigid part of the main structure, which may be either the subframe or the bodyshell itself. Because it provides a resistance to motion, the assembly must be insulated to a degree to prevent it transmitting noise, vibration and harshness to the interior. The compliance of its mounts forms a critical part of the running refinement.

5 Drive layout and its effects

Various ways of driving a car have been tried since the beginnings of motoring when the engine was squeezed in wherever there was a convenient space. The natural instinct at that time, however, was to replace the horse with an engine which pulled from the front in the same way. The complication of driving the steered wheels with the technology available forced the almost universal adoption of rear-wheel drive. It is interesting to note, in this context, that one of the first steam carriages in Europe, built by Jacques Cugnot in 1766, operated by driving the front wheels, but it was not successful. Most of the pioneering fwd systems, before 1950, came from the USA and were founded on the design of the Miller Indianapolis racing cars which first appeared in 1925.

The first recorded front-wheel-drive car of any note was revealed at the 1904 New York Auto Show by J. Walter Christie. It had a transverse V4 engine with integral final drive and independent, sliding-pillar front suspension. Christie's designs proved very successful in racing and at one time held the unofficial lap record on the banked Indianapolis racetrack.

Ben Gregory of Kansas built a series of front-wheel-drive prototypes in 1919, and some of his ideas were adopted by Harry Miller for his racing cars which dominated American racing from 1925 until World War 2. In 1920, the UK firm of Alvis built a racing car with a supercharged, straight-eight engine in a front-wheel-drive chassis for the then current Grand Prix formula, which was followed by a production sports car with the same drive system. At the same time, J. A. Gregoire built the first front-wheel-drive Tracta in France.

Independent front suspension, when it arrived, only compounded the drive-shaft problems and demanded something considerably better than the universal joints available at the time. These were large, heavy, limited in their range of movement and prone to break up if abused. Some adventurous pioneers fought back against convention because they were convinced of the logic of front-wheel drive, but in the main they had little understanding of the dynamic characteristics of what they were doing, with the result that most early fwd systems were poorly executed.

Dreams of easy-to-drive, high-performance, front-wheel-drive production cars did not die in conflict with the engineering problems involved, however. While the Americans followed Miller's lead with sports cars like the 1929 Cord, European designers took off in a distinctly different direction. In 1931, DKW introduced two front-wheel-drive mini cars, and a year later Adler launched the 1.5-litre, front-wheel-drive Trumpf, Audi came out with two front-wheel-drive models and André Citroën produced his first *traction avant*.

The balance between fwd and rwd looked set to stay, with most of the post-war

It took Ford nearly 20 years to decide to follow the Mini lead. When the Fiesta eventually came, it took fwd design several stages further than the Mini with power outputs as high as 115 bhp per tonne in the XR2 version

car industry steering well clear of fwd and all the reliability problems it introduced. There were prototypes by the score, of course, but no decision-making management team issued design approval to a new front-wheel-drive system until the genius of Alec Issigonis caused something of a design revolution with the 1959 Mini.

At the time, there were lots of reasons why fwd was right for the Mini and would have been wrong for the Ford 105E Anglia, launched the same year with rwd. Ford took a long, hard look at the Mini, rejected everything about its engineering, but kept thoughts of fwd simmering as a concept on the back burner. Certainly, there were no Mini influences in the first Ford fwd car, the Cardinal project, which was designed in the USA for production there and in Germany but was launched only in Germany as the 1962 Taunus 12M. It was not until Fiat had shown Ford what fwd could really do for a small car, with the 127, that the Fiesta concept was given the go-ahead.

Between the Mini and the Fiesta there were Renaults, Panhards, DKWs, Saabs and Lancias, all with various front-wheel-drive systems, although by 1972 less than

five per cent of the cars available on the British market had front-wheel drive. By 1982, more than 70 per cent of the cars sold in Britain used front-wheel drive, and by 1984 practically the entire US industry had switched to front-wheel-drive designs, as had the bulk of Japanese production.

Every new car launched from then on, with one or two notable exceptions such as the Ford Sierra and Mercedes 190, featured not just front-wheel drive but a total system that proved very effectively how good fwd engineering had become. In the market, more and more buyers seemed to understand and appreciate the well-publicized advantages of extra interior space and greater stability in slippery conditions.

Yet there were several major discrepancies in philosophy between models designed to compete for precisely the same section of the market, which makes the whole question worthy of detailed investigation. Why did BMW and Mercedes declare a permanent dedication to rear-wheel-drive cars? And why, when the Ascona and Cavalier became so much more successful in fwd form, did Ford stay with rwd for their Sierra?

There are some very good reasons why there is a significant and widespread shift to fwd now more than in the past, which go well beyond the dynamic aspects of front- and rear-wheel-drive concepts. Some of the conclusions are not those expected.

Design requirements in different market sectors

Minicar	*Small car*	*Family car*	*Luxury car*	*Performance saloon*	*Sports car/Coupé*	*Rally or track car*
Versatile accommodation	Versatile accommodation	Versatile accommodation	Distinctive style	Fast appearance	Sleek styling	Priority to function
Small frontal area	Small frontal area	Low C_d and reasonable area	Low C_d	Low C_d	Low C_d	Ground effect and low C_d
Small engine	Small engine	Choice of engines with diesel option	Wide choice of engines, optional diesel, possibly with turbo	Large engine with fuel injection and/or turbo	Large engine with fuel injection and/or turbo	Maximum power output
Good performance	Good performance	Good performance	Smooth performance	Fun to drive	Fun to drive	Quick responses; ultimate handling
Maximum fuel economy	Good fuel economy	Good fuel economy	Good fuel economy for class	Good fuel economy for class	Performance first	Performance first
Low cost	Low cost	Low cost	Value for money	Value for money	Cost secondary	Cost no object
Ride secondary	Adequate ride	Good ride	Good ride	Good handling	Good handling	Maximum roadholding
Easy to service and repair	Easy to service and repair	Easy to service and repair				Fast repairs at pit or service stops
Minimum weight	Minimum weight	Minimum weight	Controlled weight	Good power-to-weight ratio	Good power-to-weight ratio	Low weight
Maximum package for size	Maximum package for size	Maximum package for size	Maximum package for size	Reasonable luggage room	Reasonable luggage room	To carry long-range fuel tank and large-section spare tyre
Some noise acceptable	Reasonable noise	Low noise	Very quiet interior	Quiet at high speed	Quiet at high speed	Noise not important
4 seats	4 seats	5 seats	5 seats	'2 + 2'	'2 + 2'	2 seats only
← FWD →		← FWD or RWD →			← RWD →	

RIGHT *Market forces have really polarized the choice of drive system into two opposite arrangements with the cross-over point open to different interpretation, depending on individual priorities. In general the application of fwd is moving up-market*

Concept studies

Most car designers believe that the drive layout of any new car should be the best available solution to the total design requirement. To act responsibly in today's energy-conscious environment means designing the most efficient vehicle in each market segment, as defined by the product-planning analysts. Therefore, priorities for large cars are different from those of small cars, and as the packaging gets

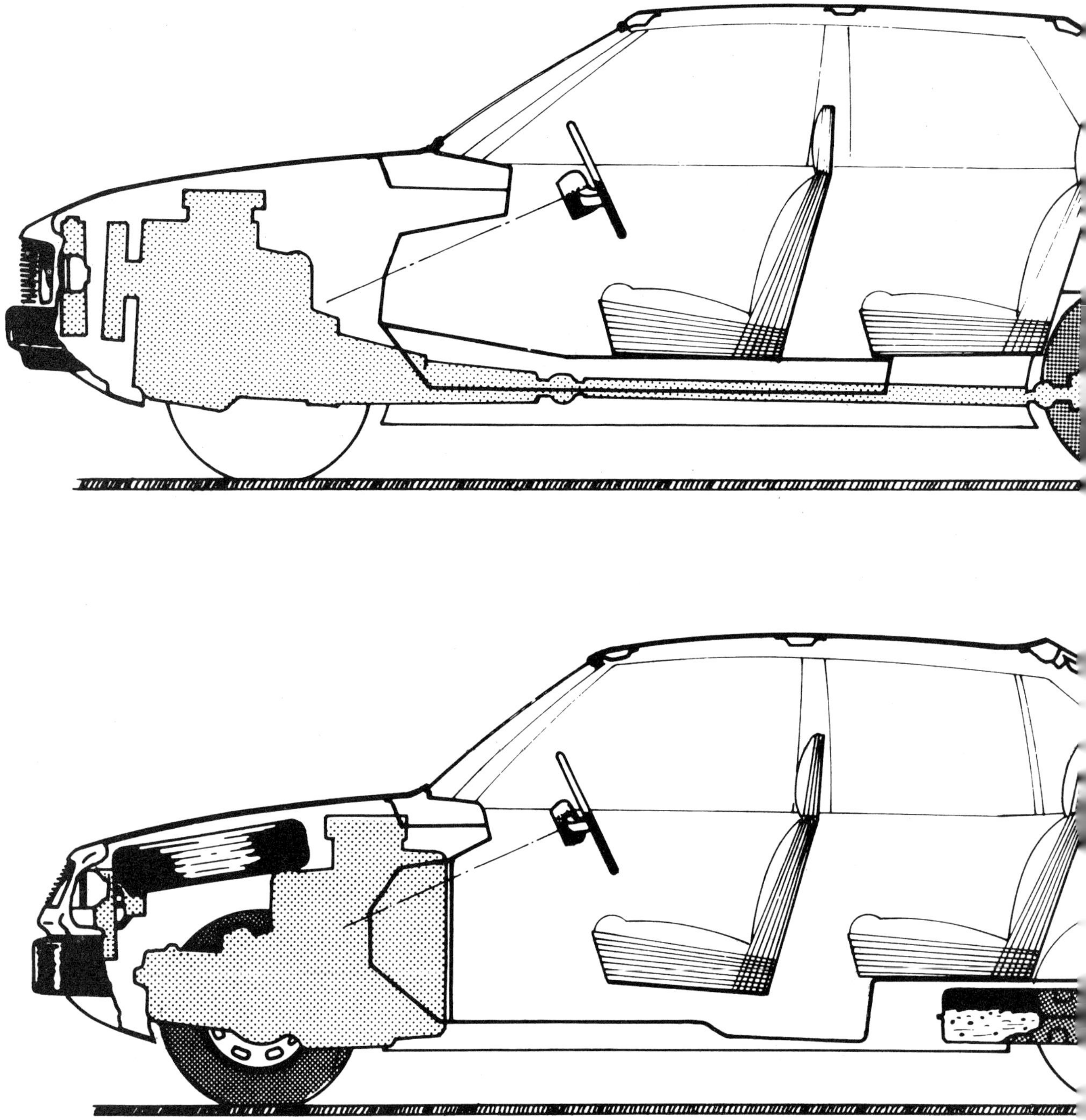

The differences in package efficiency and overall length that result are shown clearly in this comparison prepared by Renault. In addition to the whole seating arrangement moving forward with fwd, the rear end is much more roomy

'tighter', so the attractions of fwd against rwd change to produce a cross-over effect from one system to the other (as shown in the table on page 63). This view would appear to be supported by the Austin Rover Group which, after a long history of rear-wheel-drive models and a mixed range for many years, now builds nothing but fwd. Within the industry, there is no sign of unanimity about basic concepts or the specific point at which the cross-over, if any, occurs.

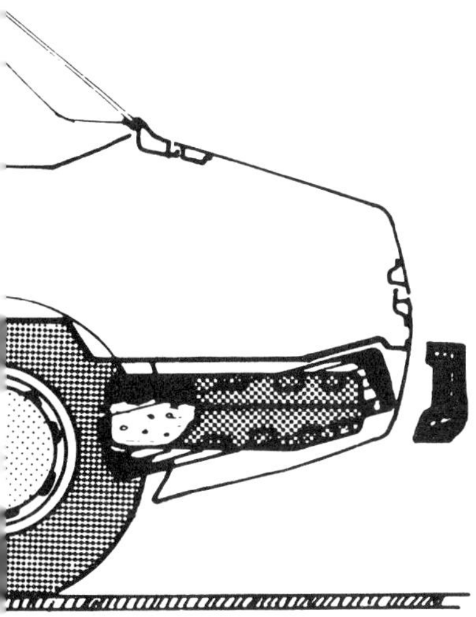

With some notable exceptions like Audi and Rover, who have been influenced radically by their co-operative technical liaison with Honda, there is little dispute among the various engineers that, for the best design solutions with today's technology, small cars up to 110–120 bhp should be fwd and large luxury cars should be rwd. In their obsession with what they call 'driving quality', BMW believe that the steering must be independent of drive influences, which produces some positive and finely-balanced steering systems but in their smallest car, the 3-series, severely compromises the accommodation provided. Volkswagen and Audi, like Saab, have proved that it is quite feasible to produce excellent, high-powered, fwd designs, although for the ultimate Audi coupé they resorted to something other than fwd.

The way the design requirements cause a shift of emphasis between fwd and rwd according to car type is fully supported by a study of power against weight for about 80 typical, middle-range, European cars. Each of the considerations affecting the design process of reaching a concept decision is worth examining in detail.

Powertrain installation

These days, it is very rare for a new vehicle design to include a new powertrain as well. New product design and development costs are so high, and the certification procedures in world markets are so complex, that both the investment and engineering work load must be spread over longer and longer periods. In the majority of cases, this means that engines are carried over from one model to the next, which can limit freedom of concept choice.

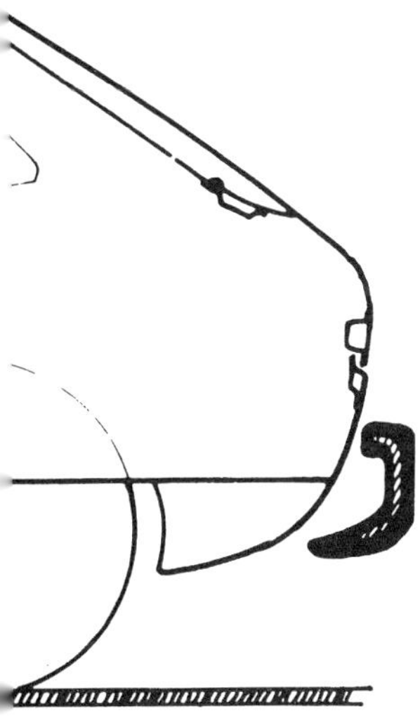

A far greater restriction is caused by the need to spread the costs over the largest sales volume possible, which leads to wider market segments for each new car. Consequently, the design must often allow for not just a single engine family but two, or even three, distinctly different types of power unit. Ford had that problem on the Sierra range of three engine families spanning 1.6 to 2.8 litres. For this spread, which stretched in customer taste from a basic fleet model to a 150 mph (241 km/h) supercar, it was necessary to install quite a large V6 as well as a small inline, four-cylinder engine.

Audi addressed the problem in their familiar, innovative way, when the 100 model 'grew up' in market terms, by introducing first a five-cylinder engine, then a turbocharger because the fwd design simply could not accept an inline six for weight and package reasons. Saab also chose the turbocharger route to more power for many of the same reasons, not the least of which was the front-end weight bias of fwd and its many dynamic effects (see below).

Aerodynamics

On the face of it, there seems no obvious relationship between the drive concept and the reductions in aerodynamic drag achievable, given a clean-sheet approach to the design. Certainly, it is possible to develop a very-low-drag shape with either fwd (as

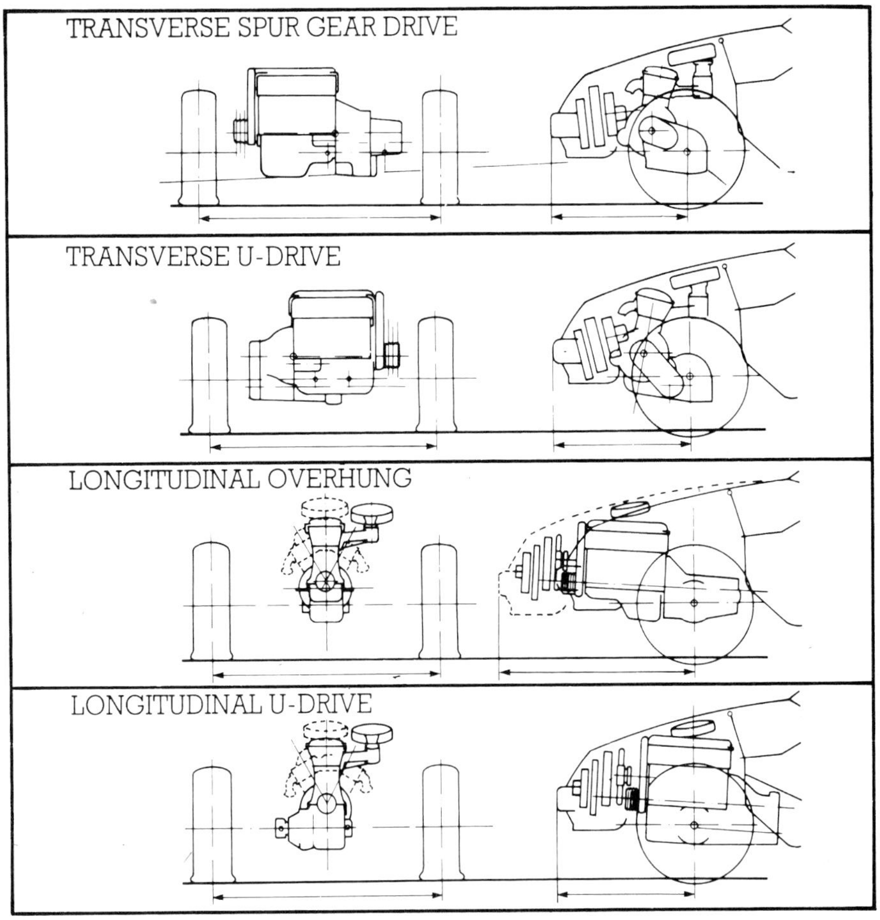

There are basically four different arrangements for engine installation and drive in fwd layouts. Each results in a different front-end package with significant aerodynamic effects caused by the front overhang and front track dimension required

on the Audi 100 with a drag coefficient—Cd—of 0.30 and the Renault 25 with 0.28) or rwd (as on the Opel Omega with less than 0.28). However, total drag is the product of coefficient and frontal area, and rwd makes the low-drag engineering task much easier.

Starting with the engine itself, length is the most critical factor. Height is only secondary, although the angle of cylinder inclination is more restricted in an fwd design with a transverse installation. BMW, for example, would find it much harder to incline their engines by 30 degrees, as they do with rwd, in an fwd layout, so their bonnet line would probably need to be raised. That means either a longer nose to keep the same slope angle or a drag handicap.

The length of the power unit is important because it pushes the bonnet line up at the front when mounted longitudinally ahead of the wheel centres, or pushes the strut housings out to create a wider car when mounted transversely. Both these effects increase the total drag, as little as 3 in. (75 mm) extra width adding about five per cent to the CdA, which would totally cancel out a reduction in Cd from 0.40 to 0.38, for example. A longitudinal installation behind the wheel centres encroaches severely on passenger room, as on the first-generation Renault 5 (rectified by a change to transverse installation in 1985).

Therefore, it is easier to design a low-drag body with a large engine if the engine can be mounted longitudinally between the front wheels and as low as possible. This does not preclude low-drag shapes with fwd, but that route (as the Audi 100 demonstrates) requires a painstaking approach to small details like cooling air flow, bonnet sealing, headlamp and bumper radii and greater control of production tolerances for consistent results.

An additional benefit of not passing the drive out to the front wheels is that the lower ends of the struts can be attached below the hubs rather than above or behind them, for better suspension geometry and a lower top-housing position.

Package efficiency

The basis for any new car design is the envelope surrounding the powertrain and suspension components, the occupants and the luggage compartment. The factor which has the greatest effect on the limits of this envelope is the drive configuration, because it sets both the toe-board and the rear-seat-back positions. Front-wheel drive usually allows the overall length to be less for the same amount of interior room, in other words greater package efficiency, especially in an effectively-planned, transverse engine installation.

Some of the reduced length is caused by shortening the front overhang, although crash protection against frontal impacts and the need to provide a gently sloping nose for low aerodynamic drag may work against this, as in the case of the Citroën CX. With the engine mounted longitudinally, the drive configuration saves no length compared with rear-wheel drive, and in cases like the Audi, where the engine overhangs the front wheels, the length is increased.

Therefore, most of the fwd saving is between the pedal cluster and the rear bumper, because the reduced size of the floor tunnel, which need cover only the exhaust, brake pipes and handbrake cables, allows the pedals to be offset and moved forward. Front wheel arches intrude more, however, pushing the pedals over towards the centre of the car and often producing a noticeable offset in the driving position.

The rear of the floorpan under the seat, which would otherwise be swept up to clear the final-drive unit, becomes available for the fuel tank or spare wheel. The suspension and location required for a non-driven rear axle can be arranged more easily into a compact, space-saving design. Therefore, while maintaining the capacity of the luggage compartment, length and rear overhang can be saved, leading to the square-backed look of the Mini, Fiesta, VW Polo, Lancia Y10 and almost every other small family hatchback. Again, though, it may not be quite as easy to achieve a low-drag body with this shape.

On average, front-wheel-drive cars are 12 in. (30 cm) shorter than equivalent rear-wheel-drive cars, which means less structural weight and a significant knock-on benefit in the size of the suspension components, brakes, wheels and tyres.

Weight and mass distribution

At first glance, and particularly for small cars, there might appear to be other significant weight benefits with front-wheel drive. According to studies carried out by BMW on a typical 2000 lb (900 kg) car, elimination of the propeller shaft and its universal joints saves about 20 lb (9 kg), while the shorter body described above saves a further 25–30 lb (11–13 kg). However, the extra proportion of front-end weight affects the steering effort and braking balance to the point where, even for medium-sized cars, power steering, a brake servo and a load-sensitive brake apportioning valve may be required. To maintain a luxury ride quality on certain models, some kind of levelling device may also be essential at the rear. The weight of these additional components could easily cancel out the weight saving achieved.

XON 126V

The car that really started the shift to fwd design in small cars was the 1959 BMC Mini, still in production nearly 30 years later. It effectively demonstrated what a compact transverse powertrain could do for a two-box package

The regrouping of components at the front in an fwd design increases the proportion of sprung mass at this end from about 52 per cent, in the driver-only condition, to about 62 per cent, which adds 20 per cent to the steering effort. Providing adequate wheel travel for a smooth ride under all laden conditions also becomes more difficult, as the ratio of fully-laden to unladen axle loading at the rear increases from 1.6:1 with rear-wheel drive to 1.8:1 or worse with front-wheel drive.

On level ground, with the driver only on board, the extra static weight of a front-wheel-drive system over the driving wheels helps traction. Uneven torque and wheelspin can be countered easily before the car twitches out of line, unlike rear-wheel drive where there is bound to be some fishtailing. This is the main reason why many drivers prefer fwd in slippery conditions.

While front-wheel-drive cars have a better static mass distribution for traction, the rearward transfer of dynamic mass, which occurs as soon as torque is applied, reduces the tractive effort available at the front wheels. Adding extra passengers or trying to tow a caravan or trailer similarly reduces the static traction advantage and compounds the dynamic effects considerably.

With a coefficient of friction below 0.6, which is typical of any wet or icy road, the relative traction available with fwd or rwd depends entirely on the load being carried; fwd provides better traction only when carrying a light load. In soft snow there is the additional advantage that the front wheels of an fwd car may be able to bite in as they drive, while with rwd they must be pushed from behind, building up wedges of snow under their leading edges.

Dynamic characteristics

When the mass transfer effects are related to the friction required to generate cornering forces in the tyres, the strong power-on understeer inherent in all powerful front-wheel-drive cars results. In extreme conditions, there is a marked secondary influence on stability as well.

As understeer is generated, the natural inclination of the driver is to apply more lock to maintain the chosen line through a bend. Should the driving torque suddenly be reduced, due to lifting off for a hazard, for example, there will be a sudden increase in cornering power which creates the classic front-end, tuck-in characteristic so common in fwd cars. With rwd, the dynamic decrease in rear-axle loading in the same situation can cause a loss of tyre grip and oversteer, although the latest generation of independent rear-suspension systems seems able to eliminate this characteristic completely.

Wheels	Tyre Usage	FWD	RWD
Front	Braking	20%	17%
	Steering	24%	24%
	Traction	24%	0%
	Cornering	17%	15%
		85%	56%
Rear	Braking	5%	8%
	Traction	0%	24%
	Cornering	10%	12%
		15%	44%
Total		100%	100%

LEFT *Tyre-usage distribution is far more balanced with rwd, as this table shows. With fwd, the front tyres are required to work over five times harder (85:15) than the rears, while with rwd the ratio is much more even (56:44)*

A fully-laden fwd car shows a greater difference in behaviour from its unladen

The car that started the 4 × 4 revolution was the Audi Quattro in its first coupé form. It soon became unbeatable in international rallies and led to a whole generation of competitive 4 × 4 rally specials. Audi now offers four-wheel drive as an option across its complete range

condition and is more difficult to control than a similar rwd car, because the changes in axle loadings are greater with fwd. Tyre loadings are similarly less balanced with fwd, giving rise to uneven wear rates between front and rear, the former sometimes wearing as much as three times more rapidly than the latter.

BMW studies also show that fwd cars as a group have a stability advantage in crosswinds, particularly if the driver holds the wheel tightly, as he is prone to do in these conditions. However, the variations between cars in each group are greater than the differences between the two systems, which confirms that stability has to be built into the basic design by development of the aerodynamic lift and suspension characteristics. Although it is probably easier to make fwd cars more consistently stable in the unladen condition, it becomes much more difficult when they are fully laden or towing.

Equally challenging, from the design standpoint, is the need to absorb all the driving forces of the wheels of an fwd system through the engine/transmission mountings. Soft mounts are required to suppress engine and driveline vibrations, while hard mounts are needed to prevent drive wind-up. In an rwd layout, these two requirements are well separated and can be designed for individually. The problems with fwd are compounded in the transverse installation by the often

The advantages of four-wheel-drive in steady-state cornering are clearly shown on these two plots of steering-angle increase (understeer) against lateral acceleration (cornering force)

Steady-state cornering (constant speed on circular course)

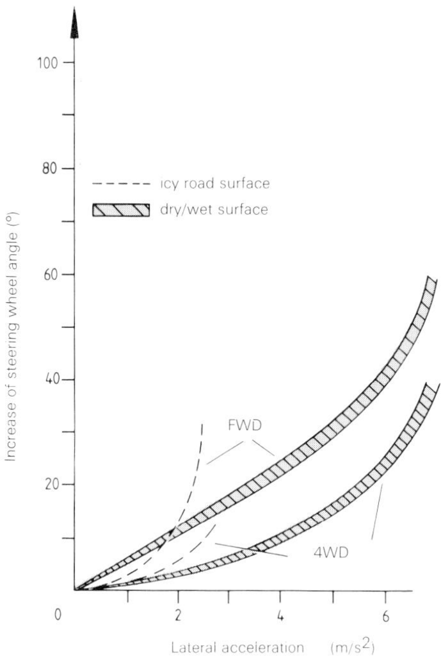

restricted room for engine movement on its mounts. Many manufacturers of fwd cars have been forced to add hydraulic dampers or tie bars (which provide unwanted acoustic paths to the structure) to engine mounts in attempts to strike an effective balance between these conflicting requirements.

An often-related problem with a transverse fwd design is that engine movement introduces bending moments into the exhaust system, which can shorten its life and make it difficult to arrange on properly-insulated hangers. With a conventional, longitudinal design, engine rocking produces only small torsional forces in the exhaust system, low-amplitude vertical shake being the only source of bending.

Unbalanced front-end mass with fwd is aggravated by mass transfer during braking, especially when unladen. Therefore, the front brakes must have a much greater capacity than the rears, and in most cases a proportioning valve to prevent rear-wheel lock in extreme conditions is essential. The availability of cheap, ventilated front discs combined with modern pad materials seldom makes front braking performance a problem, but the legal requirement for the system to operate with reasonable efficiency in the event of a partial failure has led to the introduction of diagonally-split systems on most fwd cars, which adds to the cost slightly.

When it comes to really high performance, the mid-engined rwd car has a distinct advantage on the road and the track, hence its total domination in racing. On loose rally stages, the responses required by the driver are totally different and success with fwd has only been achieved in spite of its concept, not because of it. In this kind of event, the driver needs a car which is 'nervous' in the extreme, not to the point of being unstable, but able to change direction in a flash at any time, even when sliding, and that is one thing fwd cars cannot do at all well.

Of course, all these apparent limitations of front-wheel-drive design can be solved by conventional engineering methods, as the abundance of safe, effective and quite powerful front-wheel-drive cars available today demonstrates. At the

Known as the Circle of Forces, friction theory states that the total grip of the tyre contact patch (A) can only reach a certain fixed vector limit (a). If a cornering force (C) is applied, traction (T) or braking (F) forces will be reduced proportionally. With less grip (b), the vector diameter is smaller

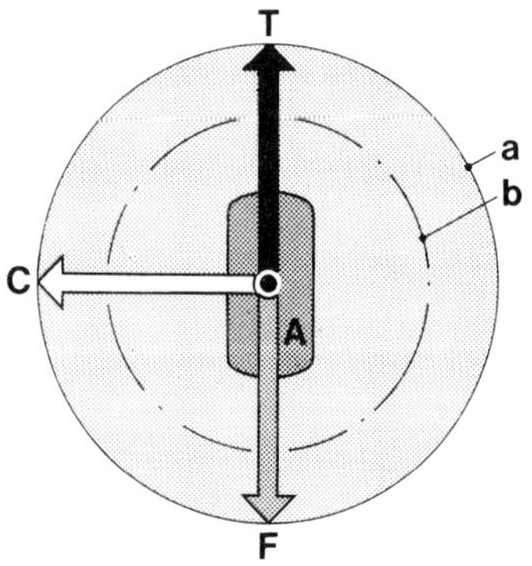

small-car end of the market, they have total domination and prove beyond doubt that the concept is effective and efficient.

At the upper end of the market, where carrying capacity and a consistent, high degree of comfort are more important, rwd is generally the only way to go. In between, it is the effectiveness of the design, and particularly its execution, which decides how good or otherwise each concept turns out to be. However, regardless of the arguments raised for either system, there seems to be something fundamentally wrong in unbalancing a car by making one pair of wheels do more than half the work. It is to the credit of their designers that this fact is totally hidden in many of the latest fwd cars, and even some of the better developed earlier ones.

Four-wheel drive

Basically, there are two types of four-wheel drive, designed for two distinctly different purposes. There are part-time, off-road systems, developed in many various forms from the concept of the famous American wartime Jeep, and there are the permanent, all-weather, on-road and rally car systems which have been spawned by the various successes of the Range Rover and Audi Quattro.

Four-wheel-drive vehicles today offer a wide choice of models, which range in character from the little, spartan Suzuki SJ410 to the ultimate luxury of the five-seater Audi 200 Turbo. In total, there are over 60 different designs of road-legal, 4 × 4 cars in production around the world.

The first main thrust in four-wheel-drive development came in 1918 when various US government departments tried to lay down a specification for a go-anywhere, military vehicle. Many different prototypes were built, from stripped-down Ford Model Ts to special tracked machines, but the task of designing something that suited all the transport and fighting needs clouded the issue too much, so no decisions were made for several years.

Using a 1932 Austin Seven saloon as the starting point, the US Infantry School developed a working set of plans for a low-silhouette, cross-country gun carrier, with an overall height of less than 34 in. (86 cm), that could perform like a 'snake in the grass'. It had balloon tyres and no suspension, which caused it to be dubbed the 'Belly Flopper'.

What eventually became the specification for the famous World War 2 Jeep was derived by amalgamating some of the Belly Flopper features with a more conventional, light car—again based on the smallest machine available in the States at the time, the licence-built Austin Seven which was called the Bantam. Following an extended military appraisal in 1940, four-wheel drive with a two-speed transfer gear was added and the key dimensions set.

Invitations to tender for production were issued to 135 manufacturers, including Willys, Ford and the American Bantam Company, who were first to submit a prototype for government test. The first contract went to Bantam, then to Willys and Ford, who were both able to copy much of the original Bantam design. Because Willys was the company with the biggest order, the vehicle was known as the General Purpose Willys—GPW for short, which soon became GP or 'Jeep'.

With the war well under way, original orders for hundreds of Jeeps were expanded to thousands from all three suppliers. In the end, the total number produced during hostilities ran to well over 600,000. Wherever the American troops went, Jeeps went with them. Their effectiveness and appearance became a legend, and many were left behind when hostilities ceased. The performance of the

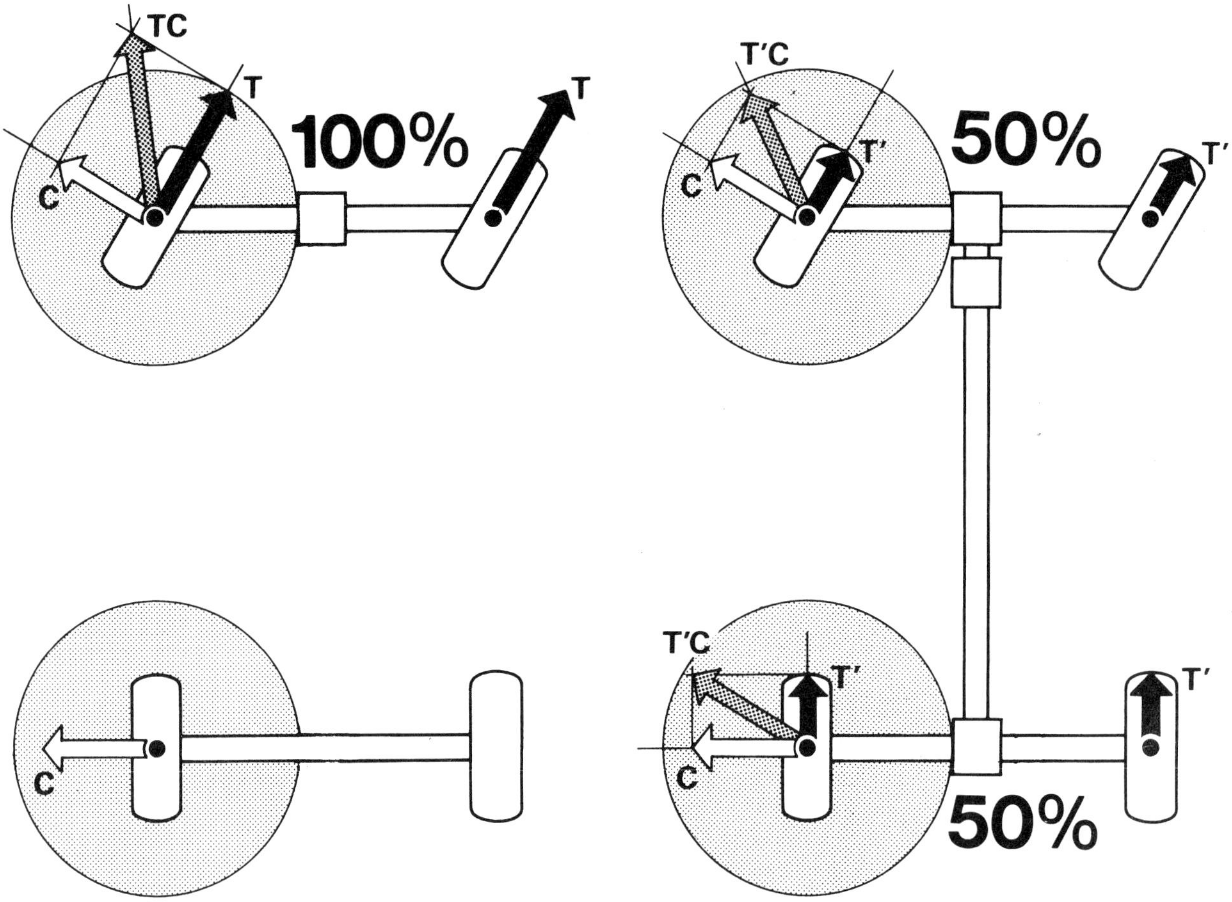

With four-wheel drive, tractive effort (T) is split between front and rear axles ($T' = \frac{T}{2}$), allowing the resultant (T'C) to reach a limit almost twice as high as with fwd (TC)

mechanical orphans of battle created a powerful Jeep influence on future off-road design in many parts of the globe.

Nowhere did this demonstration of a basic, reliable, cross-country vehicle have more effect than in England just after the war. In the blitzed Midlands car factories, a team of engineers struggled to restore prosperity to an industry equipped with spare capacity, little steel, pre-war technology and a desperate need to export. They proposed a more sophisticated, agricultural version of the Jeep, panelled in aluminium to resist corrosion, save weight and overcome the steel shortage. Using war-surplus Jeep parts for a prototype, they designed and developed the Land Rover in only 12 months. It went on sale in 1948 and was eagerly grabbed by buyers all over the world.

In other parts of Europe, all-terrain vehicle design took a slightly different direction, the emphasis being split between more specialized 4 × 4 military machines and two-wheel-drive beach cars. By the mid-1970s, however, world markets had developed to the point where almost every major producer had an entry—Ford Broncos, Dodge Ramchargers, Chevrolet Blazers, International Scouts and Jeep Wagoneers in the USA, Toyota Landcruisers, Nissan Patrols, Daihatsu Fourtraks and Suzuki Jimmys in Japan, the Fiat Campagnola in Italy, the Steyr-Daimler-Puch Gelandewagen from Austria, the Lada Niva from the USSR and the VW Iltis from Germany.

There was a well-defined, military ruggedness to all these vehicles, deliberately intended to project the macho image of the original Jeep with all that it had stood for in the war. In 1980, Audi saw the four-wheel-drive phenomenon in a different light. They took the VW Iltis drivetrain, adapted it for permanent four-wheel drive and put it in a high-powered sports coupé. Thus, the Audi Quattro was born.

In the field of private car design, the real four-wheel-drive pioneers were undoubtedly Freddie Dixon (of pre-war racing Riley fame) and Tony Rolt (pre-war ERA owner, entrant and driver). They had plans for a modest 4 × 4 family car well before 1939, but the war delayed construction of the first prototype. It was completed as soon as possible after the war and powered by a 1.5-litre Riley engine. Because the rear wheels were steered as well as the front ones, to avoid the need for a central differential, it was known as 'The Crab'. It proved unstable to drive and they ran short of money, but Harry Ferguson (the lightweight tractor pioneer who had recently won a multi-million-dollar lawsuit against Henry Ford) saw the potential and backed their efforts.

After a very long and involved development beset with delays, a four-wheel-drive Ferguson family saloon was successfully completed in 1966. It never reached production, but the same Ferguson principles were adopted on the high-performance Jensen FF which appeared in that year. The FF was a stretched version of the Interceptor made in limited numbers at a premium price. In three years, only 318 were made.

The Audi Quattro came at a time when the market was much more prepared to accept the concept, and it was backed by the full resources of the Volkswagen-Audi group. Its success in international rallies developed not only a strong consumer following but an extension of four-wheel-drive principles to the point where now every Audi is offered with optional all-wheel drive.

In 1980, not even Audi knew what a landslide in car design they were starting when they launched the Quattro. While their competitors just watched, bemused by the novelty, VAG got on with the job of developing more 4 × 4 designs. When the rest of the industry realized how seriously the buying public was taking four-wheel drive, there was little time to react. But with a good measure of help from Harry Ferguson Research, whose founder had dreamed long and hard of the day when every car would have four-wheel drive, a whole host of new designs arrived suddenly in 1985.

In a period of nine months, over 20 new 4 × 4 cars appeared, 11 of them almost being dumped on the unsuspecting public at the September show in Frankfurt. There were rear-engined 4 × 4s (like the Porsche 959), front-engined 4 × 4s (like the VW Golf Synchro and Mazda MX-03) and mid-engined 4 × 4s (like the Nissan MID-4 and MG EX-E). Everybody had one, but nobody seemed quite sure how many they could sell. Some were basic, part-time Japanese designs with fast engagement of the extra drive axle (there is a push-button on the gearlever knob for the Nissan Prairie system, and automatic selection by kickdown on the latest Subarus). Others, like Mercedes-Benz, use complicated computer controls to decide which wheels need to be driven and when. In between came the drive-on-demand Golf Synchro system that can be considered as permanent 4 × 4, although strictly speaking it is an automatic, continuously-variable selection system.

In practical terms, there are only two reasons why two-wheel drive may not be sufficient. One is the problem of driving in low-friction conditions, such as mud, ice or packed snow, when the tractive effort needed to move off may be too much for one axle to transmit, with or without a self-locking differential. The other is the difficulty of transmitting power to the road safely in a high-performance car. When there are deficiencies in a 4 × 2 system, the cross-over between the first and the second parameters begins to show up in everyday driving.

The device that is changing the whole face of four-wheel drive is the viscous coupling, developed by Ferguson and produced by GKN. As the vanes spin in the special silicone fluid, it progressively locks to transmit more and more drive

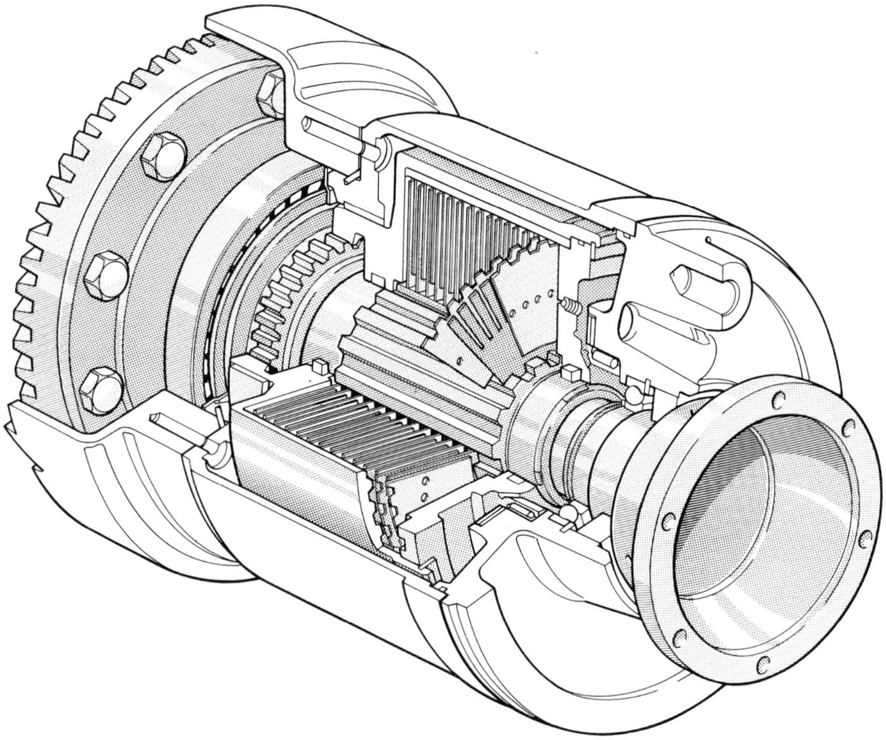

Audi reached the cross-over point sooner than most because their designs use an overhanging front engine combined with front-wheel drive. As it turned out, 4 × 4 conversion was both easy and effective for them. Looking at what others had to do to achieve the same kind of improvement only brings the whole question of 4 × 4 into dispute. How can a Koenig-tuned, twin-turbo Ferrari Testarossa lay 710 bhp on the road through two racing slicks when Austin Rover feel a 4 × 4 chassis is essential for 250 bhp in their glamorous, experimental, road-going MG EX-E coupé? Why offer 4 × 4 traction in the 90 bhp Golf only and not the GTi or 16-valve version? And why is only 33 per cent of the engine torque driving the front wheels in the Nissan MID-4 with its equal axle loadings?

Four-wheel drive is very much a double-edged sword. It keeps family-saloon wheels turning in all kinds of weather for the same reason that it is the only way to win an international rally: it more or less doubles the friction forces available to drive the vehicle forward. However, it introduces some fearful complications into the drive layout in return for specialized benefits that most drivers may never even use.

It is much easier not to have four-wheel drive at all than to make it foolproof, which is why the industry has turned its back repeatedly on 4 × 4 concepts for ordinary road cars. What changed all that was the clever Ferguson viscous coupling (vc), used now by BMW, Ford, MG, Nissan, Peugeot and VW, with plenty more in the pipeline. Smoothly and progressively, it prevents all the 4 × 4 advantages being lost through one end of the car; it is compact, insensitive to wear and elegantly simple in concept. At the launch of the Golf Synchro, VW research engineers even claimed its stiffness could be adjusted simply by changing the fill level of silicone fluid, which opens the door to a computer-controlled adaptive vc in the not too distant future.

While the task of adding a self-engaging front axle to the rear-engined VW

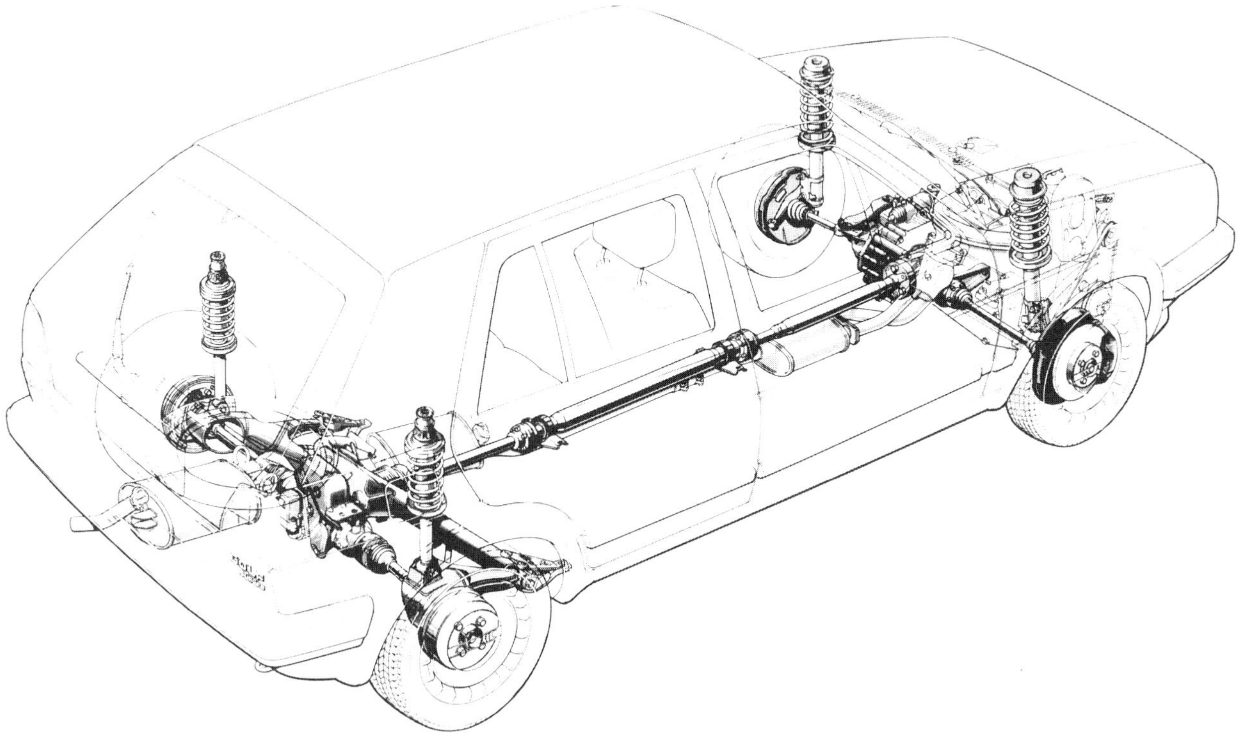

Volkswagen has extended permanent four-wheel drive into the field of small cars with the ingenious Synchro Golf that uses a central viscous coupling to distribute drive to front or rear wheels according to the grip available on the road

Transporter or Microbus introduces few handling problems, doing the opposite to the fwd Golf becomes an acute engineering challenge. Under emergency braking on slippery surfaces with the Golf, there is enough rear drag to lock-up the rear wheels, a characteristic outlawed by European braking legislation but common to every rigidly-connected 4 × 4 system. To prevent this potentially-hazardous behaviour, a neat, one-way roller clutch has been added at the rear-axle crownwheel output to make the rear drive freewheel on the overrun.

Then it was found that the car would not reverse up a slippery slope, as the clutch prevented the rear wheels being driven backwards. Adding a sliding peg-plate freewheel lock, driven by a vacuum piston under electric control from the reversing light switch, cured the problem. Next, it was found that the car could not be 'rocked' out of a bogged-down situation (there are no lateral differential locks or anti-slip devices), so a prevention device was needed to hold the freewheel lock engaged until a forward gear is selected (fortunately, gear-position sensors were already built into the transmission to work the economy gearchange indicator).

Packaging the propeller shaft and rear drive into the Golf takes away about a third of the luggage capacity, raising the floor by about 6 in. (15 cm), even with a space-saver spare wheel fitted. Had the Golf not been designed to take a blow-moulded and very convoluted plastic fuel tank, its fuel capacity would have been reduced and the range cut back as well. The central tunnel (which covers the propeller shaft but normally is only a stiffening member for the bodyshell) is higher too and over 1 in. (2.5 cm) wider (2.2 in./5.5 cm at the front end). The car is 198 lb (90 kg) heavier, adding 0.5 sec to the 0–60 mph time, with a three-per-cent, fuel-economy penalty.

Mercedes-Benz steered clear of the Ferguson patents altogether, as did Porsche who use a multi-plate clutch to engage the front axle progressively on the 959 supercar. On the W124 estate cars only, M-B 4-Matic uses a fluid-drive coupling

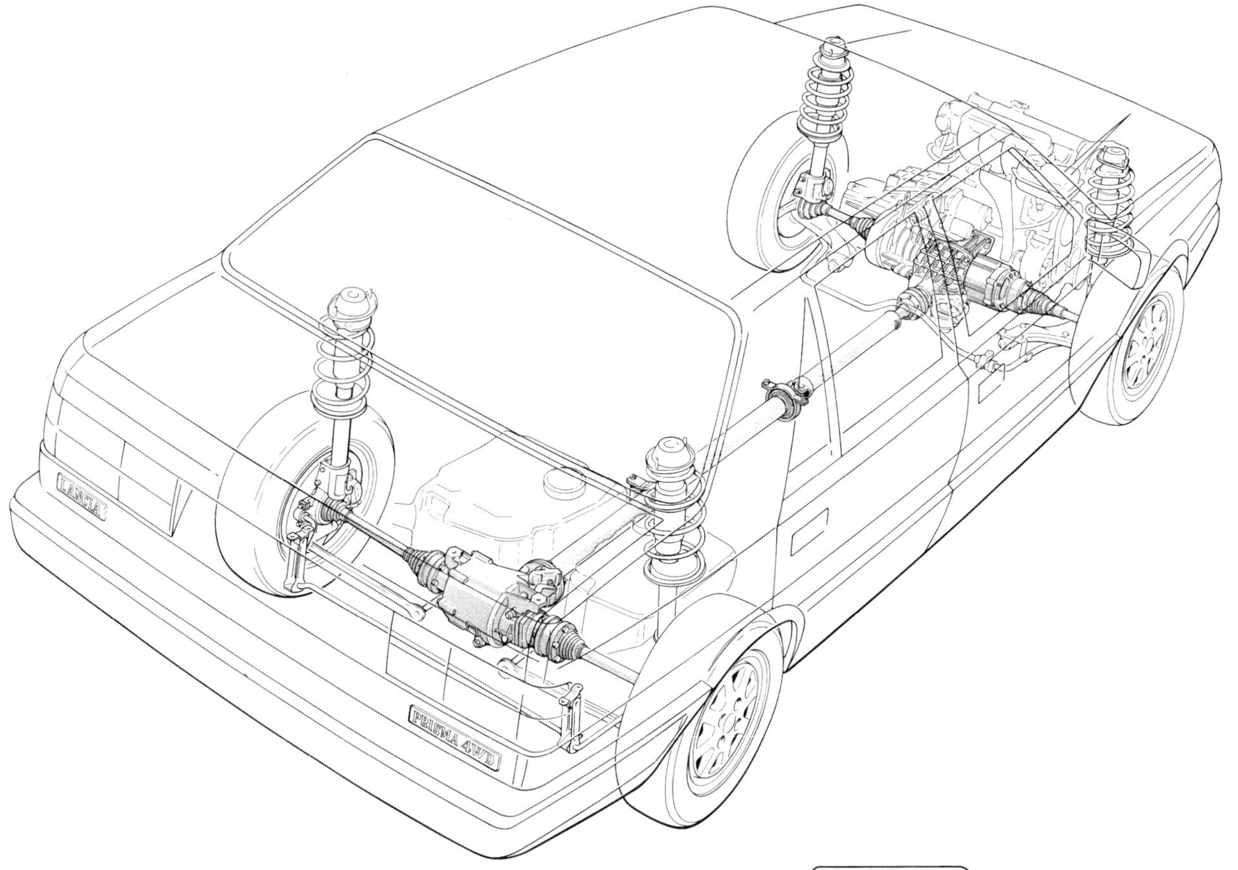
LANCIA
PRISMA 4WD

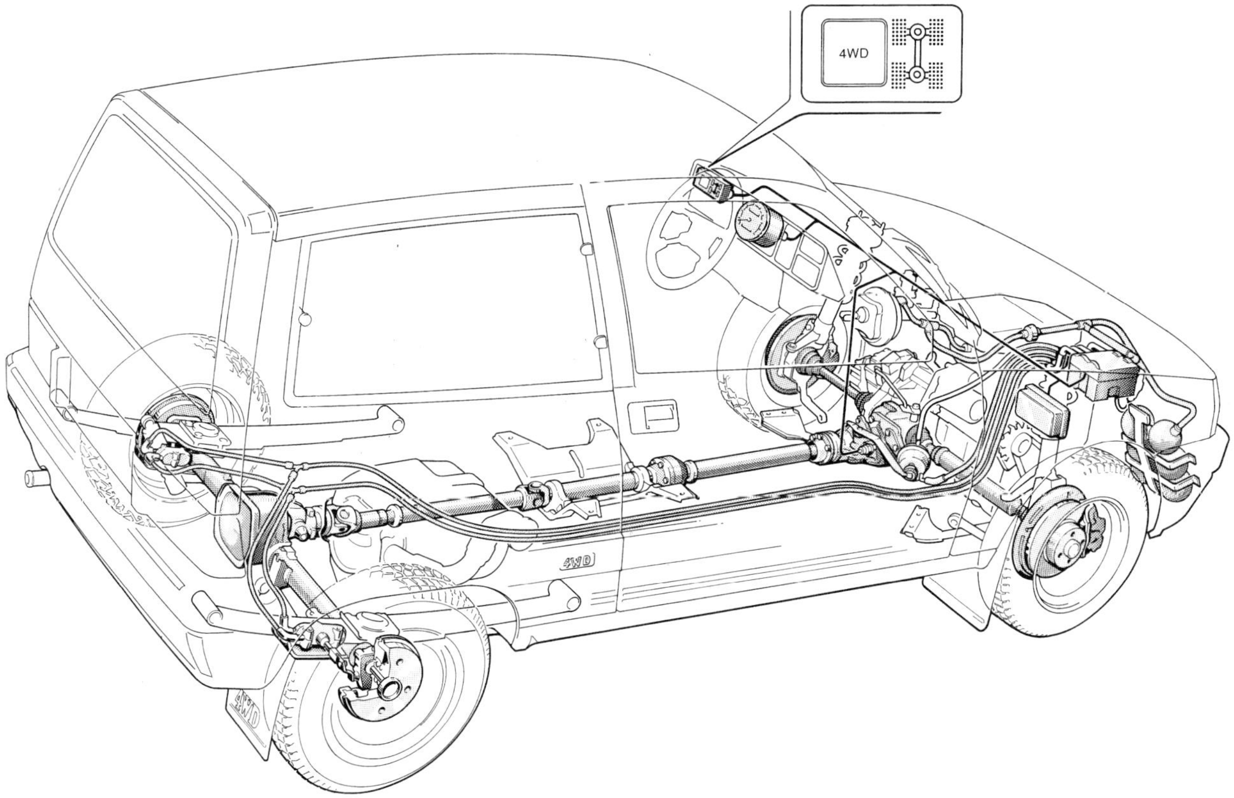
4WD

ABOVE LEFT *Lancia has three different 4 × 4 systems in production. On the Prisma there is permanent all-wheel control with a 56:44 torque split between front and rear and a manually-locked rear differential*

LEFT *On the Lancia Y10, drive to the rear wheels is engaged by the driver operating a fascia push-button. It selects rear drive and locks rear hub freewheels*

In the high-performance Lancia Delta HF there is permanent four-wheel drive to all four wheels, similar to that on the Prisma but with a self-locking rear differential

4x4 CHARACTERISTICS

Make & Model	Engine/Drive	Front/Rear Drive	Centre Diff	Front Diff	Rear Diff	Drive Loss Mode*
Alfa 33 4x4	Front/FWD	Part-time	Fixed drive	Free	Free	1 front + 1 rear wheel
Audi Quattro	Front/4WD	48/52	Manual diff lock	Free	Manual lock	1 front + 2 rear wheels
BMW 325i 4x4	Front/4WD	37/63	Diff + VC	Free	VC	1 front + 2 rear wheels
Fiat Panda	Front/FWD	Part-time	Fixed Drive	Free	Free	1 front + 1 rear wheel
Ford RS 200	Mid/4WD	50/50 or 37/63	Diff + VC	Diff + VC	Diff + VC	All 4 wheels
Ford Sierra 4x4	Front/4WD	37/63	Diff + VC	Free	VC	1 front + 2 rear wheels
Honda Shuttle	Front/FWD	Part-time	Fixed drive	Free	Free	1 front + 1 rear wheel
Mazda EM-03	Front/4WD	53/47	Free	Free	LSD	1 front + 1 rear wheel
Merc 4-Matic	Front/RWD	50/50 + coupling	Auto diff lock	Free	Auto lock	1 front + 2 rear wheels
MG EX-E	Mid/4WD	35/65	Diff + VC	LSD	LSD	All 4 wheels
Nissan Prairie	Front/FWD	Part-time	Gear drive	Free	Free	1 front + 1 rear wheel
Nissan MID-4	Mid/4WD	33/67	Diff + VC	Free	LSD	1 front + 1 rear wheel
Opel Kadett	Front/4WD	50/50 to 30/70	Diff + LSD	LSD	LSD	All 4 wheels
Porsche 959	Rear/RWD	Clutched 50/50	Diff + clutch	Free	LSD	1 rear wheel (35%)
Subaru XT	Front /FWD	Part-time	Gear drive	Free	Free	1 front + 1 rear wheel
VW Golf Synchro	Front/4WD	100/0-0/100	VC (no diff)	Free	Free	1 front + 1 rear wheel

* Drive Loss Mode describes wheels which must lose traction before car gets bogged down

Four-wheel-drive systems do not all provide additional traction to the same levels in all kinds of road conditions. Most 4 × 4 road cars will get bogged down if one side cannot find enough grip

combined with automatic locking differentials in the centre and at the rear to provide three stages of four-wheel drive under microprocessor control. Like most so-called 4 × 4 systems, including Audi's, it reverts to two-wheel drive at the limit of its performance (see the above table).

Unlike the Ferguson coupling, which detects speed differences within its own system, M-B 4-Matic requires wheel-speed sensors to take measurements and input the computer so that comparisons can be made with stored dynamic targets. If these signals show a rear wheel is about to lose grip, the centre coupling engages to drive the front axle, with all three differentials unlocked at this stage. If insufficient traction is still detected, the centre differential is locked progressively and hydraulically until it is locked fully. If that does not provide enough traction (i.e. wheelspin is detected by comparisons of the sensor signals), the rear differential locks in the same way. In this condition, the drive is just like that in a Quattro with the manual locks engaged. As soon as the brakes are applied, the 4-Matic systems switch out, allowing abs to function in the usual way.

Data stored in the computer's memory includes a relationship between steered-angle and wheel-speed differentials, so that the front drive is not engaged every time there is a slight amount of front understeer. The calibrations must have taken years to perfect, and only Mercedes-Benz has the courage and expertise to tackle the complexity of such an intricate, electro-mechanical system. However, it is the kind of technology edge Mercedes buyers expect and goes a long way towards explaining why BMW are so reticent about promoting their Ferguson-developed system for the 325i.

Nissan approached Steyr-Daimler-Puch for help in developing their mid-engined 4 × 4 coupé, which would have made a competitive Group B rally car had the regulations not been changed in the interests of safety. Drive for the front axle passes under the engine sump after leaving an epicyclic differential and passing through a Ferguson VC. Torque output of 28.4 mkp (206 lb ft) from the 3-litre, 24-valve, four-cam V6 is split 33:67 between front and rear, with a conventional, limited-slip differential within the rear axle. The MID-4 design has variable-geometry rear-wheel suspension called HICAS, which has significant effects on its high-speed handling (see Chapter 11).

6 Computer aids and design techniques

Although the first use of computers in suspension design and development dates back over 20 years, rapid and recent advances in graphic displays, measurement techniques and processing time are putting a new emphasis and value on the results of interactive studies. Today, the visual display unit linked to a powerful, high-speed computer running advanced, dynamic suspension and vehicle models are the basic tools of the suspension-design trade.

Computer aids in suspension design eliminate much of the guesswork, empirical measurement and practically all the tedious calculations that were required previously to determine safe stress levels and acceptable durability standards as the preliminary stages to a new system design. Carefully-developed programmes allow the thousands of simultaneous equations of dynamic movement to be solved rapidly, and advanced computer graphics are used regularly to simulate actual parts in action under strain, showing differences in stress levels by keyed colour graduations, like a thermal-image picture of heat radiation. Three-dimensional representations also act as valuable visual aids to the experienced eye of a component engineer working on a new design, with light-source angle of view and magnification all under keyboard control.

This kind of computer-aided design (cad) is currently being integrated with computer-aided manufacture (cam) into a new operating science known as computer-integrated engineering (cie). It is having dramatic effects on the whole process of creating new vehicles and components, and updating existing models.

Stress analysis for both the supporting elements of the integral body/chassis unit and each component part of the suspension system is performed with impressive accuracy by a mesh of finite elements (fe) built up into a framework of several thousand, finely-detailed, interlinked struts forming triangles, squares and rectangles. From the first applications of fe programmes, used to handle only discrete parts of a full system, new tools have been developed that can simulate the considerably more complex models required to represent authentic dynamic behaviour.

Early fe work was limited originally to stress and strain calculations by the inability of the computer systems to handle the non-linear events and relatively large displacements typical of vehicle suspensions under dynamic loads. Setting up representative mathematical equations of motion for the interrelated parts was too complex and time-consuming for the skills and abilities available within engineering department staffs. However, with new-generation computers came automated and user-friendly programming interfaces that eliminated the tedium and lowered the threshold of acceptance by traditional mechanical engineers.

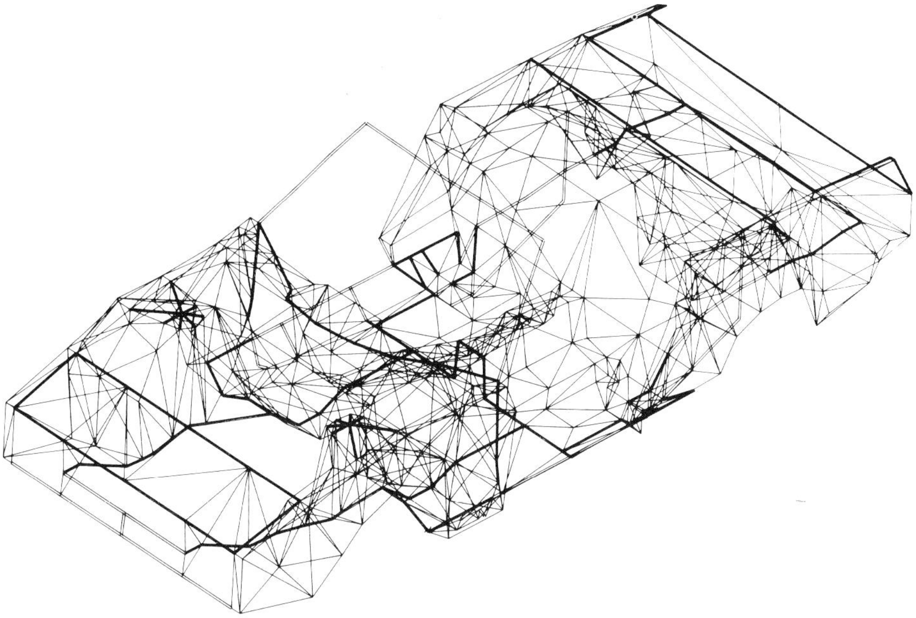

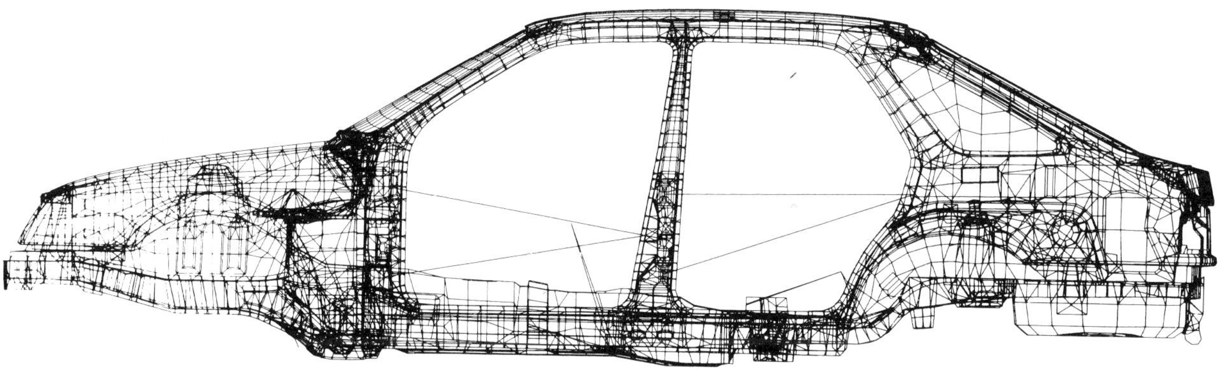

By representing the structure of the car as an integrated mesh of triangulated struts and panels, stresses in the complex assembly of body panels can be analysed and optimized

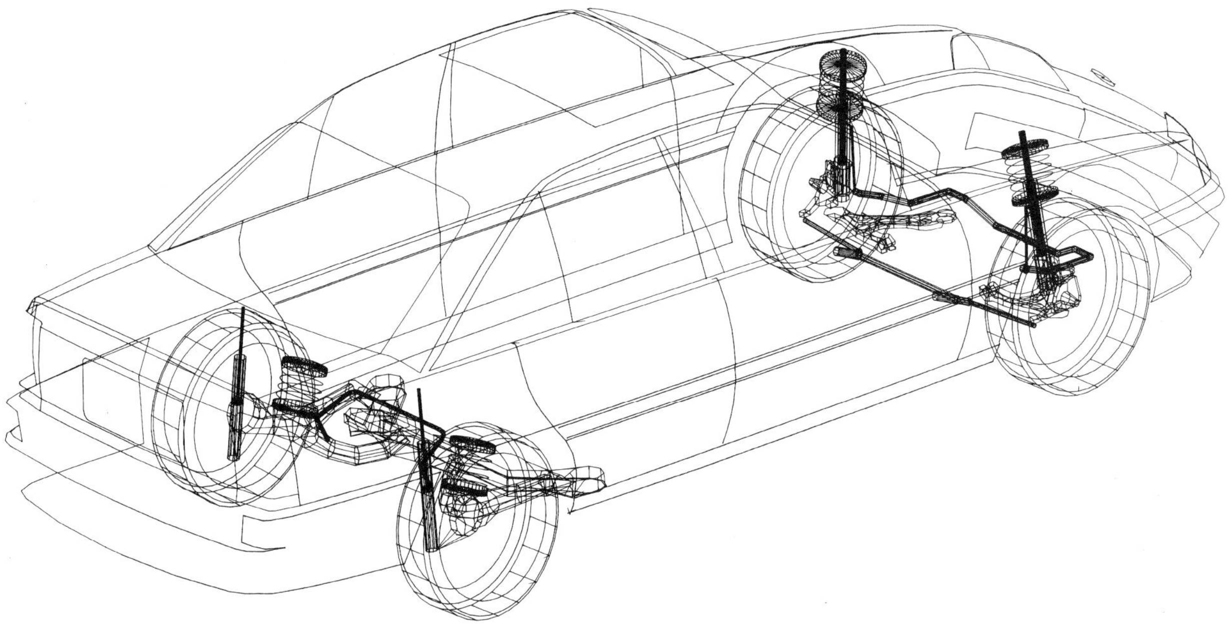

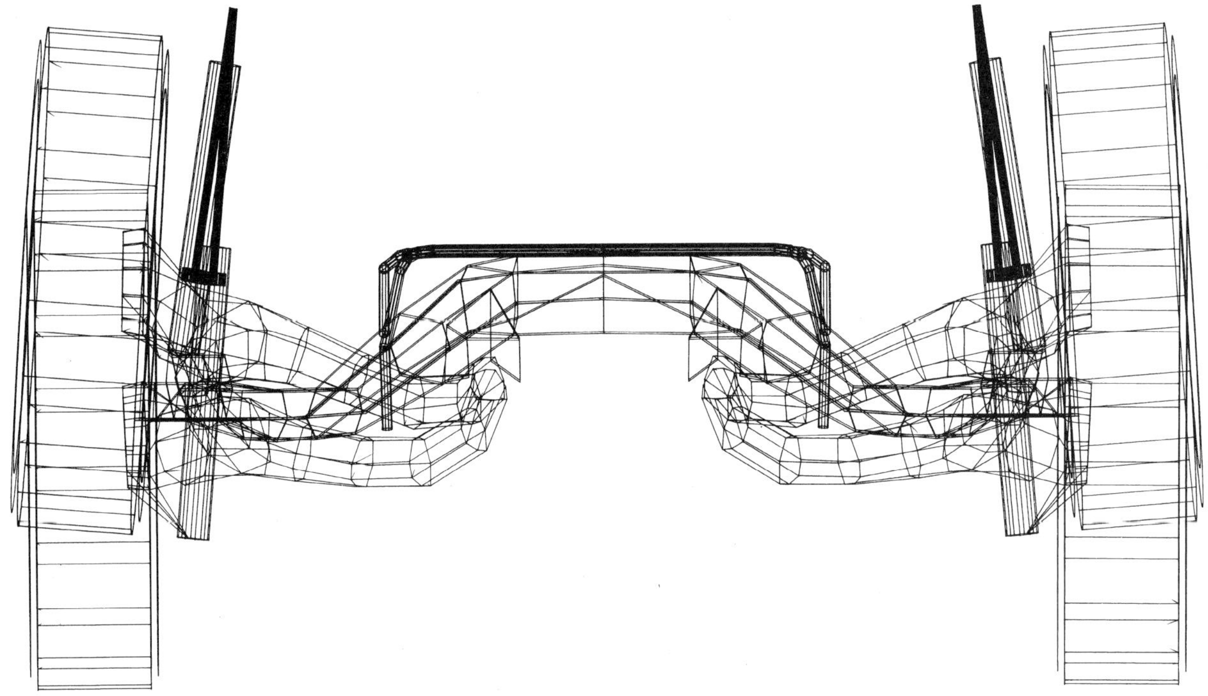

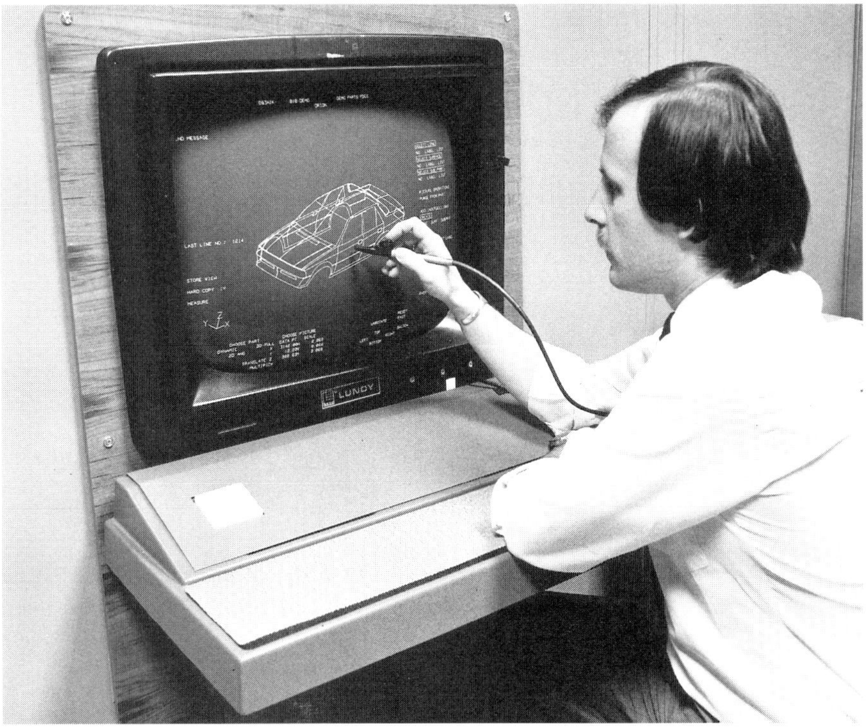

RIGHT *Computer graphics like this interactive system used at Ford are transforming the working methods used by chassis and suspension engineers*

As mentioned already, fe methods of structural analysis represent a complex component or assembly as a multiplicity of regions or elements within which simple stress (and strain) assumptions are taken to apply. Each element interacts with those adjacent to it, and these interrelationships are expressed as a system of linear, simultaneous equations containing many different unknown variables. Without the considerable processing power of modern computers, the equations would be impossible to solve.

The method was used originally by the aircraft industry and has been adopted since by shipbuilders, civil engineers and the automotive industry. It has allowed the science of stress analysis to be extended way beyond the first, two-dimensional studies into the realms of three dimensions where the computer can solve equations for each given load input to satisfy the conditions of equilibrium at common points on each element's boundaries, giving a set of displacements as the answers. From these solutions and the elemental stiffnesses, the programme works back to obtain stress levels and, if required, carries out further processing to derive the vibration characteristics.

Fe methods are such a powerful computer tool that their analytical capabilities outstrip stress engineers' knowledge of loadings and material properties in a component such as a suspension member. Using techniques known as comparative stressing, new designs can be developed for optimum shape, weight and strength from a database of previous test results.

LEFT *Finite element (fe) models on large mainframe computers are used in the design of dynamic systems in the vehicle (top) and in the analysis of separate components (bottom)*

By analysing new proposals for their section moduli (relative strength in bending), moments of inertia (relative stiffness) and cross-sectional area (relative weight), direct comparisons can be made with existing or historical designs to derive performance predictions. Instead of laboriously building co-ordinate inputs to a computer programme from manual readings of shape and section thickness, digitizing tablets are used to save time and increase accuracy.

For computer studies of suspension characteristics, the complex wheel and spring system is represented by several idealized elements with simplified but representative relationships

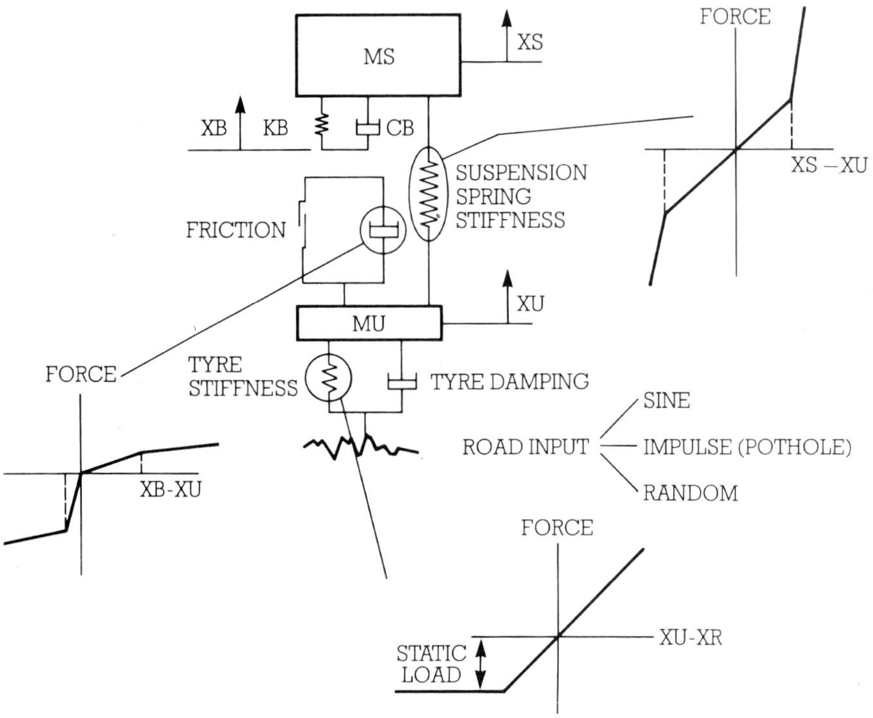

Data and specifications derived in the development processes are stored in memory devices, transferred to component suppliers on magnetic tapes and often used directly to make parts on numerical-control production machines. Where drawings (hard copies in computer parlance) are required, they are usually generated automatically at extraordinarily high speed in computer-controlled drafting machines. In many cases, data transfer between terminals is achieved by telephone or fibre-optic land-lines, satellite telecommunications links or microwave transmissions.

To appreciate the perspective of computer applications in engineering design, it is worth looking at Ford's European operations, because this organization recently completed a five-year programme to expand the use of computer aids in every aspect of product-development operations. There are now over 3500 computer terminals in use by Ford in Europe.

A special systems department is responsible for providing computer capacity and services to meet the needs of all functions within the engineering management groups. It pioneers and operates new software tools and ensures that new technology is introduced in a structured and disciplined way. The department is split into three functional divisions—Planning, Development and Services—which all support both arms of the geographically-separated but organizationally-integrated British and German engineering centres, where there are major data centres connected to a pan-European network of mini-computers and microprocessors. This network is also able to access Ford's extensive American computer centre in Dearborn via transatlantic cables and satellite links.

Computer-aided engineering staff are responsible for developing and supporting the use of computer-based analytical tools, primarily using fe techniques, while cad specialists aim more to streamline design, engineering and manufacturing processes by the introduction and development of computer graphics and numerical-control (nc) machining and manufacturing systems. Computers are also used extensively for the automation of test procedures through the introduction and maintenance of advanced component rigs and sophisticated data-collection vehicles (see Chapter 10).

Data and communications control is handled by specialists responsible for the planning, development and operation of the extensive engineering computer network. There is also a systems team specializing in office effectiveness and information, concerned with integrated office systems through the use of intelligent computer work stations and engineering information systems using the latest relational database technology.

In this typical computer model of a car and suspension system there are 6 degrees of freedom with simulation of the engine and body masses, the passengers in their seats and each unsprung tyre and wheel assembly

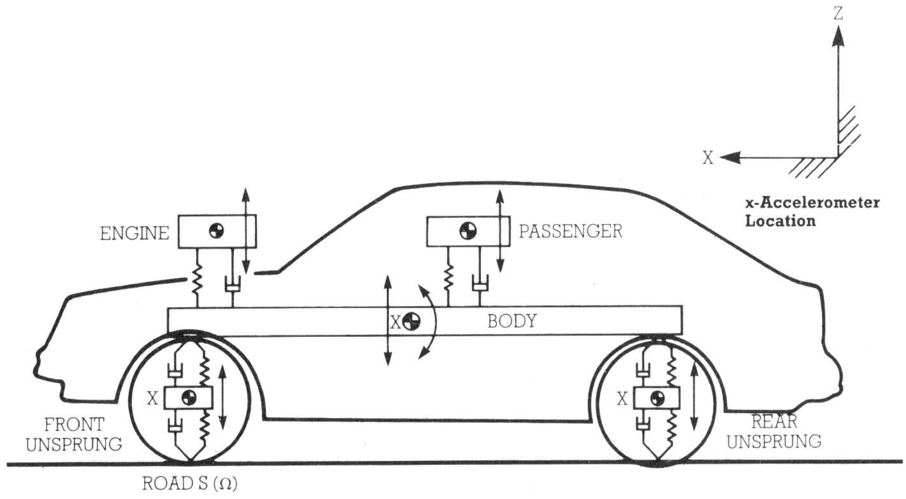

Hardware is built into an integrated European computer network of IBM-compatible mainframes, ranging from IBM 4331 through Amdahl 580 to an IBM 3081K, a CDC 730 and several DEC VAX computers with links to Europe and the USA. There are also some small UNIVAC and Burroughs systems with mini-computers from DEC, Prime, Datapoint, Computervision, General Automation and Wang, and various types of micro.

All the large car manufacturers, and even some of the smaller specialist consultants, also have a considerable number of advanced computer programmes for use in the design and analysis of ride and handling matters. They range from a single-wheel model used to predict the non-linear effects of suspension movement on vehicle ride to a sophisticated simulation of the total vehicle, including steering, brake and drive torque, bush performance and side forces.

The primary advantage of all these programmes is that they allow the first prototypes to be built closer to the optimum, which can shorten the lead time for new models and release more time for refining and developing production systems. Although originally developed for use on large, mainframe computer systems, like those housed in Ford's vast computer complex at Dearborn and the latest Cray computers used by General Motors subsidiaries like Opel in Germany, some have been rewritten for use on the latest generation of relatively-small, high-speed mini-

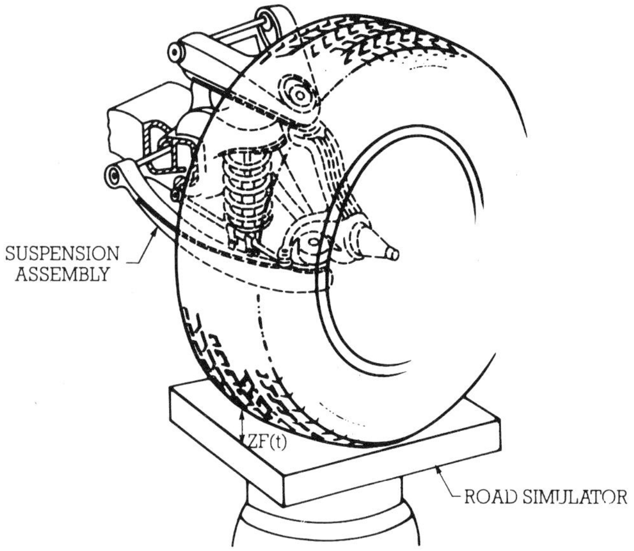

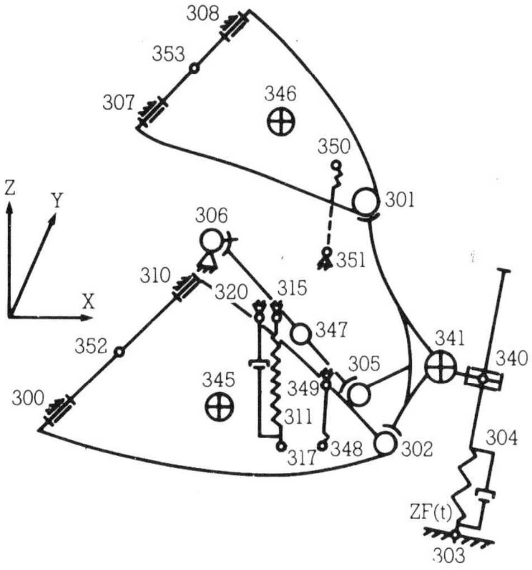

Many computer simulations (like the mathematical model of a single-wheel suspension system on the right) are based on original data gathered from instrumented components subjected to rig tests in the laboratory (like the road simulator on the left)

computers available locally throughout the industry. Some of these programmes have been developed by the technology-based, computer-software and engineering-analysis industry for sale to design departments around the world. Others have been developed in-house by the car makers themselves to suit their own needs and systems.

One of the primary dynamic programmes, developed and copyrighted by Mechanical Dynamics Inc. of Ann Arbor, Michigan, and in widespread general use all over the world, is called ADAMS (Automatic Dynamic Analysis of Mechanical Systems).

The ADAMS programme can perform ride, handling and durability analysis, both for advanced research and product development. It can take into account all the compliance-steer effects of front and rear suspensions and the characteristics of tyre flexibility, allowing designs to be refined before hardware has been built.

The origins of the ADAMS programme go back about 20 years, but only since it has been possible to incorporate adequate tyre data has the programme been of real value. It is based on a three-dimensional, multi-degree of freedom, rigid-body system which performs large amplitude displacements to predict velocities, accelerations and forces. Total vehicle handling simulations are constructed to duplicate actual vehicle kinematics, statics and dynamics, including non-linear load/deflection and force/velocity characteristics, as well as time-dependent steering inputs, brake and drive torques and external forces like side-wind gusts.

Outputs available include time-response plots of displacement velocities and accelerations of any modelled part, forces between two parts, and detailed, animated graphics that can be viewed statically or frame-by-frame through a storage terminal. Simple 'stick and circle' representations on the computer screen depict vehicle behaviour accurately under a wide variety of driving conditions, including U-turns and 360-degree circle tests at speeds up to the overturning limit, and slalom or lane-change manoeuvres up to spin-out.

Another suspension-modelling programme designed to simulate the handling characteristics of a complete vehicle has been developed by Ford and is known as the FRESH (Ford Research Handling) simulation. This system was developed to capture the vehicle behaviour necessary to study questions of dynamic response and the often intangible qualities described as 'driving feel'. Of particular interest were the characteristics of front-wheel-drive vehicles, rough-road handling, aerodynamic effects and behaviour at the limit of tyre adhesion. To simplify the procedures and calculations, vibration, noise and structural loads were not considered initially.

The structural elements of the body are represented as a rigid system, with the suspension modelled as a set of kinematic relationships that govern both the position and orientation of the wheels relative to the body. Tyre forces and moments are resolved into components at the centre of each wheel and quantified by reference to co-ordinate points fixed on the body. Both independent and solid axle systems can be modelled for all aspects of their dynamic behaviour.

Although conventional terminology is used to describe movement about the three normal axes—pitch, roll and yaw—some aspects have been developed into more complex and realistic relationships. Simple, fixed roll-centre height, for example, is replaced by a variable suspension formula for a more accurate representation of rear-end jacking in extreme manoeuvres. The processing power of the computer also allows complete dynamic equations to be used, including the higher-order terms often ignored in previous analyses.

To complete the authenticity of the model and represent the behaviour of the vehicle in transitional conditions, like braking or accelerating in a turn, the rotating motion of the wheels has also been included. Special attention was also paid to modelling the steering system in order to represent torque-steer phenomenon and

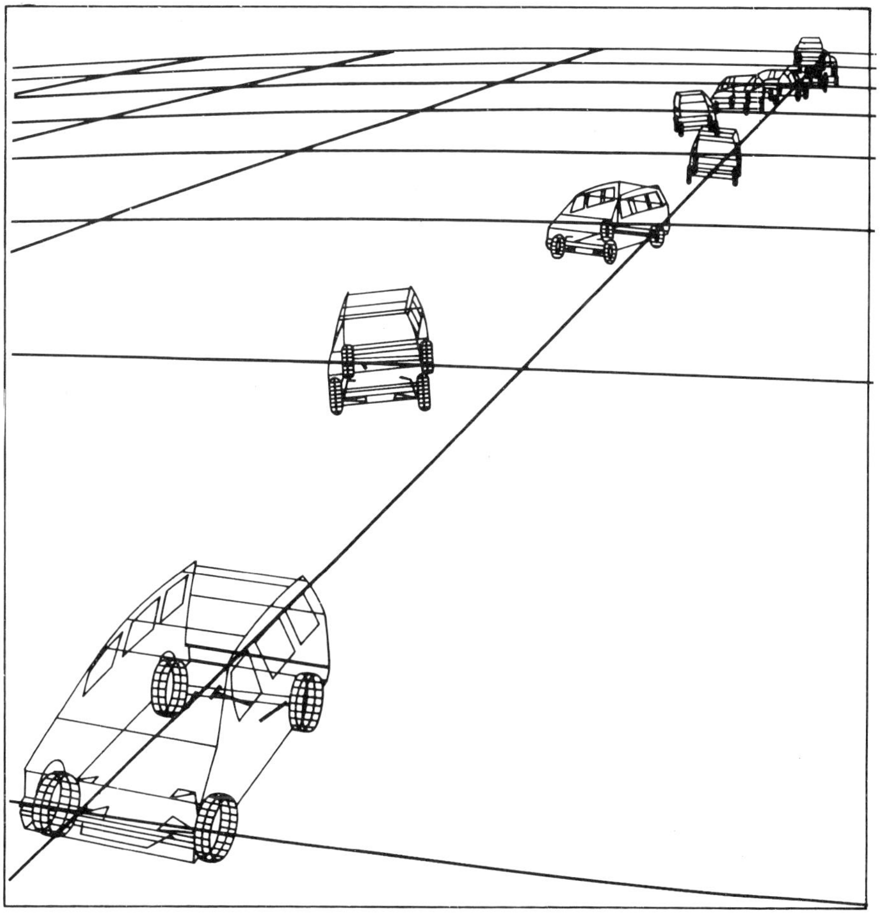

In some of the computer programmes available to suspension engineers, actual responses to inputs can be displayed as animated graphics like this slalom test for a double lane-change manoeuvre. It is used to investigate vehicle stability under extreme conditions

free-control response (hands-off stability). Details of driving torque and drive-shaft angles are also included for front-wheel-drive vehicles.

The resulting mathematical model has 17 degrees of freedom, including six modes for the sprung mass, four describing wheelspin conditions and three describing the steering system. It is defined by 30 coupled, first-order, differential equations, increased to 38 by including the forward and lateral accelerations of the wheels. Although tyre behaviour has been simplified, using previously-determined values for the forces generated by the slip angles, the programme still occupies about 7000 lines, most of which are documentation, input/output and plotting routines, and general overhead and management procedures.

FRESH simulations are capable of reproducing a wide range of operating conditions and control situations. They have been used to predict performance of different suspension systems on a particular vehicle, to study rough-road handling, to predict crosswind stability of different aerodynamic configurations and to investigate torque-steer effects in front-wheel-drive designs.

Flat-road directional response is a particular study where FRESH has proved useful. By specifying a steering input which increases slowly until the limit of capability is reached, a pseudo steady-state cornering curve of lateral acceleration against steering-wheel angle can be generated for any chosen speed. Another typical analysis predicts the rate of yaw generated by a transient 'step steer' input (a favourite road test of jerking the wheel suddenly in a swerve manoeuvre). Different suspension characteristics, including system inertia, produce markedly different transient responses which can be illustrated on a plot of yaw change rate against time.

Other applications of the FRESH programme can be used to generate response-time histories for random steering inputs and to investigate the effects of different body shapes or suspension designs on aerodynamic stability. By running models with alternative configurations through a simulated 25 mph (40 km/h), 90-degree side gust, wind-tunnel findings can be related more closely to actual driving performance.

Overall, the FRESH simulation programme is proving to be an accurate and efficient tool for predicting the dynamic behaviour of vehicles. Careful development of the structure and frequency content has reduced the computer effort while retaining the solution accuracy.

Another example of a suspension computer aid is the DRICOM (Drive/Compose) optimization mechanism, which is used to establish initial suspension geometry so that wheel envelopes can be developed to allow the package engineering of a new model or system to proceed. It runs on large, mainframe computers, like the Honeywell Multics or the Control Data 176, but can be adapted to any of the latest engineering systems, like the Cray installed by Opel.

The DRICOM programme predicts static movement and clearances required to accommodate different ride heights, steering and driveline angles, and the effects of component changes in the early stages of vehicle design.

The front-suspension element in the combined programme is known as COMPOSE (Computer-Oriented Mechanism Programmed for Optimum Suspension Effect), while the rear-suspension element is known as DRIVE (Driveline/Rear-suspension Interaction and Vehicle Effects). COMPOSE is a computer simulation of the complete front-suspension and steering mechanism of an independently-sprung, fwd vehicle. It locates all pertinent points of a symmetrical suspension in three-dimensional, vector space and analyses the relationship between these points as the system moves from full bump travel (jounce) to full rebound, and from extreme left turn to extreme right turn.

It was developed specifically to analyse front suspension with front-wheel drive, where there is complex interaction between the suspension and steering which controls the universal-joint angles completely. As a new suspension layout is designed from basic parameters and assumptions, a single database is developed from which all vehicle and component engineers can work and which can be reproduced through computer graphics as a stick model. Therefore, the clearance envelopes required to interface with engine-compartment layout, toe-board position and all the other package considerations are fixed at a very early stage.

Within these constraints, the programme can be used to develop a feasible suspension in detail and generate viable alternatives. It can accept both wishbone-and-strut location systems for a wide variety of designs with either front- or rear-wheel drive. It responds to all standard angles and relationships of king pin inclination, toe, track dimensions, castor, camber and Ackermann effect, as well as anti-lift, anti-dive and roll-centre geometries in printed and graphic form. It rejects configurations which violate the specified constraints and can analyse up to almost 60,000 variations of a given suspension.

DRIVE can be used for a wide range of suspension configurations from link-located solid axles to fully independent systems. It investigates the noise, vibration and harshness (nvh) effects of drive-shaft angles, as well as clearances for a variety of powertrain alternatives. Through DRIVE, it is possible to optimize joint angles to reduce torsional excitations and improve interior comfort levels. Following analysis with DRIVE, test vehicles have shown exceptionally good nvh characteristics.

Suspension design techniques

The design process for suspension components begins with an exchange of ideas between the management office controlling the total programme and the engineering department involved, determining the concept and establishing the functional, weight and cost objectives. These will be influenced by many factors, such as customer demand, the engineering and manufacturing capability of the company and its suppliers, the package and cost constraints, and the overall timing plan.

During development, potential failure modes and effects are forecast, and any concerns eliminated by established methods of modification. Stress analysis is studied using the latest cad systems, and data generated for automatic drafting and cam techniques. From this data, materials and parts are procured to the specification required.

Prototypes are built for exhaustive testing on special, accelerated-durability rigs and on vehicles driven over extended distances at both security-screened, company proving grounds and locations worldwide, including hot and cold climates. During this period, many special tests are devised, in association with the component suppliers, to prove the parts under development. Often, these involve techniques such as computer-controlled, vehicle-simulation durability rigs and vibration tests using stress mapping and cumulative-damage analysis of results.

Coincidental with these activities, manufacturing and purchasing feasibility are sought together with visits to suppliers to monitor progress and resolve any concerns which may arise. From then on, the engineering responsibilities are extended to include any aspects involving the production co-ordination of the finished parts, including supervision of the functional build and launch phases on the assembly lines and any service concerns during the model life.

Integration of cad/cam

At first glance, the logic and desirability of providing uninterrupted data flow from cad work station to the finished working parts might seem a simple task to achieve. When the penetration of cad systems is relatively low, there are considerable difficulties in applying computer benefits across the board. It takes a radical change in working methods, and often totally new concepts in manufacture using higher levels of automation and robotics, to reach worthwhile levels of integration.

The completion of the process, known as computer-integrated manufacturing (cim), will allow designers to respond much more quickly to market changes with greater flexibility. Current methods demand high levels of experience and expertise, especially in such complex subjects as suspension design. The cim system spreads the knowledge base much further (almost to every computer terminal) and eventually will be integrated with commercial data processing, marketing, stock control and finance.

In 1986, European manufacturing companies spent about £8000 million ($12 billion) on equipment and systems for applying cim methods across the board, from terminals to robots. By 1990, it is predicted that the annual figure will be more than double that amount. Before the technological revolution in design, development and manufacturing reaches that level, new universal standards of data-communication language, known as manufacturing automation protocol (map), will be introduced to ease the transition and compatibility of all the competing systems. General Motors have already made proposals to link different suppliers' systems to their own communications network with a map system, although the spearhead of technology leadership still lies in the aircraft industry with McDonnell Douglas in the USA.

7 Tyres and their influence

As mentioned earlier, in suspension terms, the tyre should really be considered as a secondary spring working in series with the main road spring. But the tyre is far more complex than a simple spring and plays a vital part in the dynamic handling behaviour of the vehicle.

The roots of the pneumatic tyre go back to well before the dawn of motoring, to an original patent first filed by R. W. Thompson in 1846. His tyre had a stitched, leather casing containing a number of separate, air-filled tubes. However, it was William Dunlop who set the world on its way to running on pneumatic rubber tyres when he 're-invented' the system for bicycles in 1888. A year later, when asked to repair an English tourist's glued-on Dunlop bicycle tyre, Edouard and André Michelin, makers of rubber balls and brake blocks in Clermont-Ferrand, France, decided to develop a better, removable design.

What followed has become one of the most technical and mysterious fields in automotive engineering, involving a most unusual interface between sciences, materials and concepts. As well as acting as an advanced form of pneumatic, steel-and-rubber spring, the modern tyre provides all the traction and braking required in high-performance vehicles, generating quite staggering levels of cornering force and lateral grip when stressed and distorted well away from its natural state and shape.

The elements of a tyre's construction are simple. The working medium is the air chamber which converts input loads and forces into stresses applied to the carcass, or flexible framework, to which the rubber is bonded. In the analysis of a tyre's behaviour, the factors to consider are the design of the carcass, the composition of the rubber and the pattern of the tread ribs and blocks. Each will be dealt with separately before complete tyre designs are discussed.

Carcass construction can be of two different types: cross-ply or radial-ply. In a cross-ply tyre, which was the first design to be adopted, several layers of cords are laid diagonally at alternating angles between the beads (stiff, multi-stranded wire rings moulded into the open carcass lips at each side to hold the inflated tyre against the wheel rim).

The angle the plies make with the plane of the wheel sets the balance between directional stability and ride comfort, because they control the way in which the tyre can distort. In most cross-ply tyres, the most acceptable compromise is obtained when this angle is 45 degrees, although there are some designs where it has been reduced to as little as 30 degrees for specific stability or speed requirements.

The modern radial tyre uses several layers of different materials including rayon and steel to build up a highly developed carcass on to which the synthetic rubber compound is moulded

This typical radial tyre has four diagonal rayon plies (A) above two radial plies (B) shaped into a low-profile 70-series profile (C). The rubber is moulded round the shoulders (D) from a dual-compound with deep drainage channels (E) and large tread blocks (F)

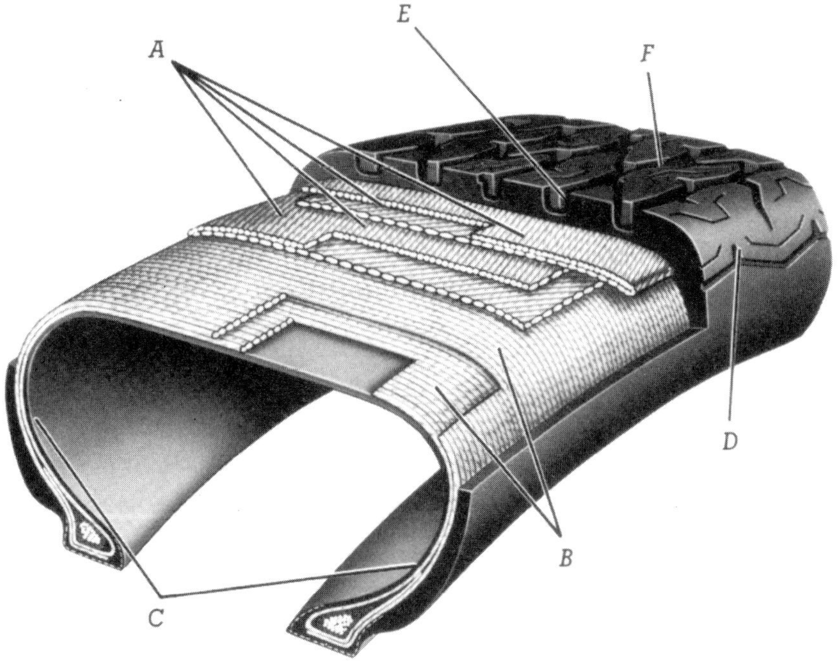

For the past 20 years, however, there has been a long-term trend towards radial-plies, or separate cord layers laid in the same direction as the wheel plane and at 90 degrees to it. This allows the two conflicting design requirements to be handled separately, although the interaction of one ply against the other can never be eliminated totally.

In addition to these two main types of tyre, there are variations which span the differences between them with a number of hybrid permutations.

In both cross-ply and radial-ply tyres, the number of layers and the material from which they are made are the factors which decide how the tyre will perform. The plies can be of flexible fabrics like rayon, nylon and other members of the polyester family, or more rigid materials such as fibreglass, steel or graphite.

The main physical characteristic that has the greatest effect on ride is the sidewall stiffness, although road inputs also contain a horizontal vector derived from the forward motion of the wheel. Soft sidewalls, on the other hand, delay the response between the wheel and the tread footprint, usually leading to unstable handling in extreme conditions.

The mechanism of the rolling tyre is complicated and not always as predictable as might be expected or desired. As the distortion caused by the static mass of the vehicle passes along the tread, a compression wave is formed across the contact patch, causing a shuffling and abrasive action on the rubber in contact with the road. When this wave is distorted by bumps, tractive effort from the driving wheels or braking forces, the situation becomes very confused indeed. Cornering loads and steering inputs add further to the pattern of distortion and the complexities created.

It would be extremely difficult and inappropriate to enter into all the details of modern tyre design. That part of chassis and suspension engineering is best left to those involved in the tyre industry, where there are many well-versed and experienced specialists. Suffice to say, at this point, that all carcass designs have individual characteristics which need to be allowed for in the development of the complete vehicle.

Rubber compounds and tread patterns

In ultimate handling terms, probably more significant than the design of a tyre's carcass is the nature of the compound used for the surface of the tread. Although the laws of physics governing sliding motion state quite clearly that frictional forces (F) can never exceed vertical loading (R), a rolling tyre does not behave in a theoretical manner. As long as the wheel keeps turning, friction coefficients ($\mu = \frac{F}{R}$) will be greater than 1.0 at the limit, especially on a coarsely-dressed surface. Studies into this behaviour show that the rubber deforms to match the macro features of the surface to generate a kind of rack-and-pinion effect.

By virtue of its origins, natural rubber is a very gummy substance. Modern tyre treads are made from a mixture of natural and man-made materials which are blended together to form what is called a polymer.

When several molecules of a material combine together to form a larger, more complex molecule (often referred to as a long molecule or macro-molecule) with the same empirical formula as the simple molecule, the process is known as polymerization and the resultant structure is a polymer. Various polymers of the same material can be formed by chemical reactions that cause the basic chemical building blocks to join together in different ways.

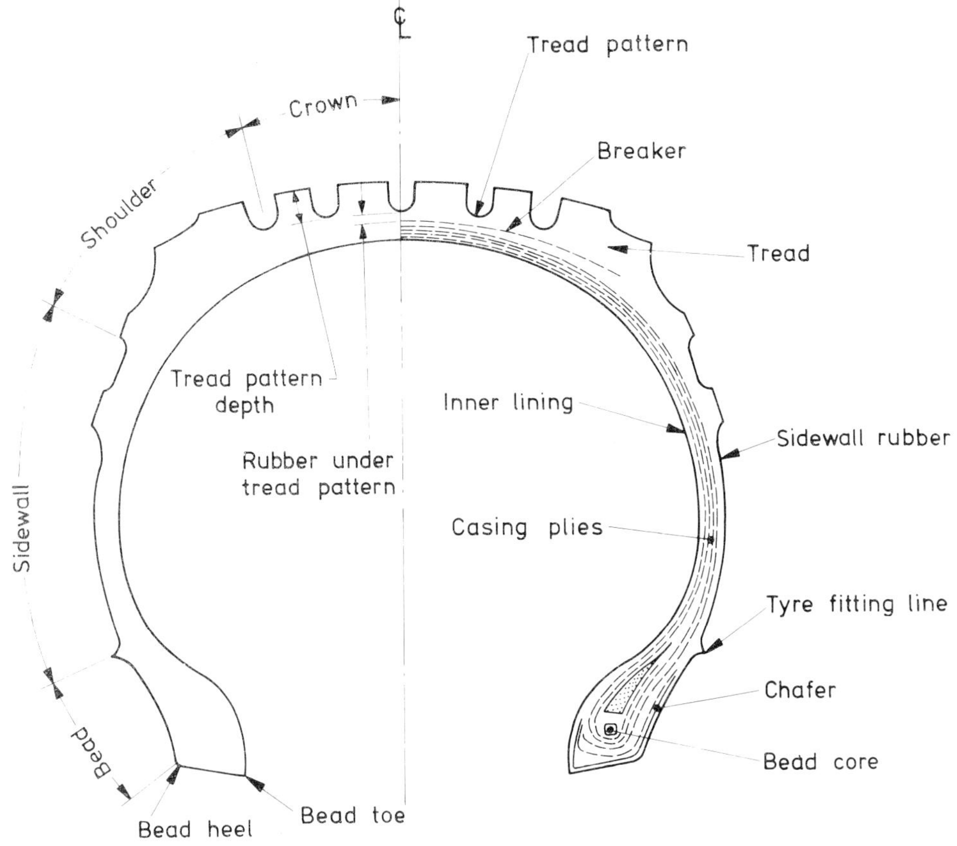
CL
Tread pattern
Crown
Breaker
Shoulder
Tread
Tread pattern depth
Inner lining
Sidewall rubber
Rubber under tread pattern
Sidewall
Casing plies
Tyre fitting line
Chafer
Bead
Bead core
Bead heel
Bead toe

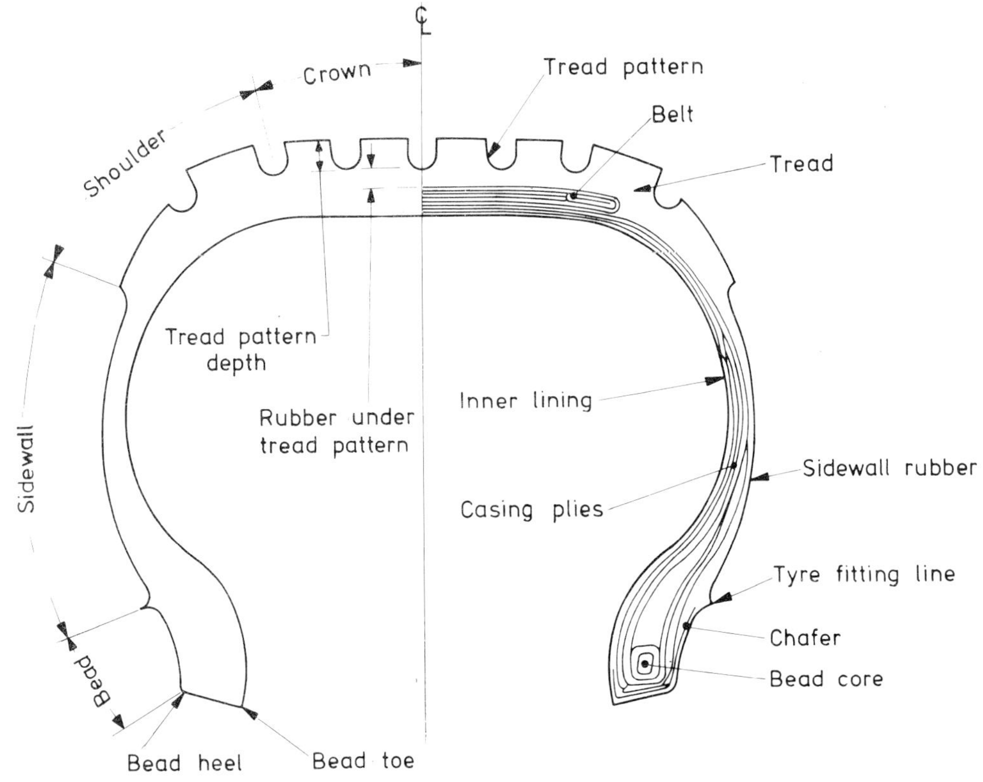
CL
Crown
Tread pattern
Belt
Shoulder
Tread
Tread pattern depth
Inner lining
Rubber under tread pattern
Sidewall
Sidewall rubber
Casing plies
Tyre fitting line
Chafer
Bead
Bead core
Bead heel
Bead toe

The difference between cross-ply (top) and radial tyre construction (bottom) is shown in these two diagrams of carcass construction. Each part of the carcass has a specific name

There are two major classes of polymer—plastics and elastomers (with different physical properties, as the names suggest)—while some polymers can also be formed into sheets of film or strands of synthetic fibres.

Plastics are formally defined as a large and varied group of materials that consist of, or contain as an essential ingredient, a substance of high molecular weight which (while solid in the finished state), at some stage in its manufacture, is soft enough to be formed into various shapes—usually through the application of heat and pressure. These materials are sub-divided into two distinct classes according to their fundamental behavioural characteristics.

One class, known as thermoplastics, become soft when exposed to heat and harden when cooled, regardless of how often they are subjected to this cycle. That means they do not work-harden or become brittle in any way, allowing them to be reshaped many times over. The other class, known as thermosetting plastics or thermosets, are set into a permanent shape by heat. Once set, they cannot be remelted or returned to their original state.

Elastomers are really synthetic rubbers and, like rubber, display various degrees of elastic properties. They are similar to plastics only insofar as they are composed of long, molecular polymer chains. Unlike plastics, which have a rigid or crystalline structure, they have an essentially mobile (or colloidal) structure of molecules. They cannot be moulded under pressure alone, as they behave elastically and try to regain their original shape when loadings are relaxed. However, elastomers can be moulded, with difficulty, into forms that do retain their shape and which normally allow considerable levels of distortion without plastic yielding or fracture.

There is no sharp dividing line between plastics and elastomers. In fact, the two categories are so close that an intermediate group of polymers, known as thermoplastic elastomers, has been developed recently.

Both categories of plastics and elastomers can be sub-divided further into materials commercially available in bulk as general-purpose commodities, and materials of a more specialized engineering nature where properties are tailored to the operating environment or load conditions.

Man-made (synthetic) fibres form the basis of most modern clothing materials in the general-commodity bracket. In the speciality group, one class is known as the aramids, of which Kevlar is probably the most famous. Fibres are used to reinforce various plastics (such as epoxy resin), usually in the thermosetting category.

From that slight diversion into the field of chemical engineering, it can be seen that almost every part of a tyre can be tailored to provide the best engineering compromise for its performance requirements. What matters to the suspension engineer is how well the tread compound stands up to wear, how well it absorbs small-amplitude vibrations and how it behaves as it approaches the ultimate limits of its frictional grip.

Excessive heat is the major enemy of materials like elastomers, but their characteristics also change at low temperatures and they can be seriously degraded by acids, chlorine, hydrogen, sulphur, oxides of nitrogen, and even long-term exposure to atmospheric oxygen. The tyre designer is committed to making sure there is plenty of useful life in every tyre long before any of these effects becomes a matter for concern.

Today's tread patterns have come a long way from the simple, circumferential grooves of 30 years ago. The prime function of the tread is to provide drainage for water. On a clean, dry road, the smooth slicks of a racing car place the most compound on the surface and give the best grip. When there is a film of water

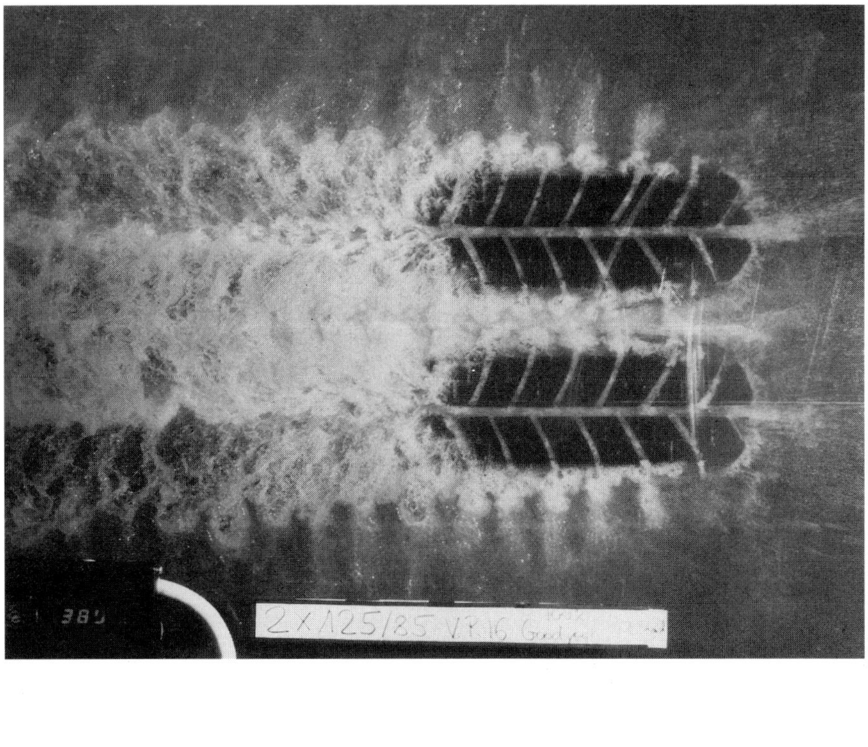
2 X 125/85 VR 16

Juhan twin tyres appear to offer many dynamic and safety-related advantages, but require special wheels. They have been made experimentally by Goodyear and are under development by Avon

between the two, the tread must clear it away so that at least some of the compound can come into contact with a relatively dry part of the surface. The ability to disperse water rapidly is a critical feature in tyre-tread design.

If there is too much water for the tyre to cope with and an unbroken film remains, what starts as a simple lubricating layer builds up, allowing a wedge of water like a standing wave to force itself under the tyre and cause all control of the vehicle to be lost. Eventually, all tyres will reach this stage, known as aquaplaning, if the water is deep enough or the speed is too high for the tread to provide effective drainage.

Other considerations in tread design are mostly concerned with the noise generated by the wind resistance of each separate block and the acoustic effects of their rhythm as they beat against the road. Computers are used now to design patterns which are randomly spaced to avoid natural resonances.

Although most tyres follow the standard concept of fitting between the raised flanges of a wheel rim, there have been recent moves towards the incorporation of an inner safety ledge on the wheel, which is designed to prevent the tyre pulling off the rim after a sudden deflation. An example is the joint development of the TD system by Michelin and Dunlop (now SP Tyres). There have also been one or two more ambitious new projects, among them the use of twin tyres.

Twin tyres and run-flat systems

Tyres mounted in dual formation have been used on the rear axles of commercial vehicles for over 50 years, providing a simple and cheap means of increasing the load capacity, with a degree of extra safety in the event of a sudden deflation. Twin rear wheels also appeared on some pre-war competition cars to improve traction in short sprints.

However, the use of double tyres on single rims, front and rear, for cars is a relatively new and controversial idea. Although there is an obvious and simple logic to the concept, the first public showing of an engineered system by Czechoslovakian-born, Geneva-based inventor Jerry Juhan (at the 1983 Frankfurt Motor Show) caused considerable reaction among the established European tyre industry.

The benefits of twin tyres in action are clearly shown by this picture taken from underneath a glass plate at 60 mph. The central gutter between the tyres is very effective in clearing away water

About five years earlier, Dunlop had tried fitting two slim, cross-ply motorcycle tyres to a Jaguar to assess the potential of such a system, but their test results were not encouraging and they abandoned the idea quite quickly. Consequently, the feelings about the JJD wheel concept (as Jerry Juhan's patent called it) were generally very mixed.

The rationale behind twin, narrow-section tyres is quite simple. If a very large drainage channel is cut around the centre tread of a high-performance, 50-series tyre, it dramatically reduces the possibility of aquaplaning on deep, standing water. Separate the two halves to form their own air chambers for safety, and you have two relatively-narrow tyres mounted side by side. These are easier and cheaper to make. More importantly, the system offers a valuable and unique back-up against the dynamic effects of a sudden blow-out at speed, and provides the means to continue the journey with minimum disruption.

From a commerical point of view, the JJD system had significant connotations. First, it offered a valuable safety bonus that was easily perceived and appreciated by the customer, who would obviously be prepared to pay a premium for the extra security. Second, it would allow the expensive and heavy (but little used) spare

This section through the special Juhan wheels shows how the two tyres are mounted independently with provision for the inner valve to be extended to the outer wheel surface

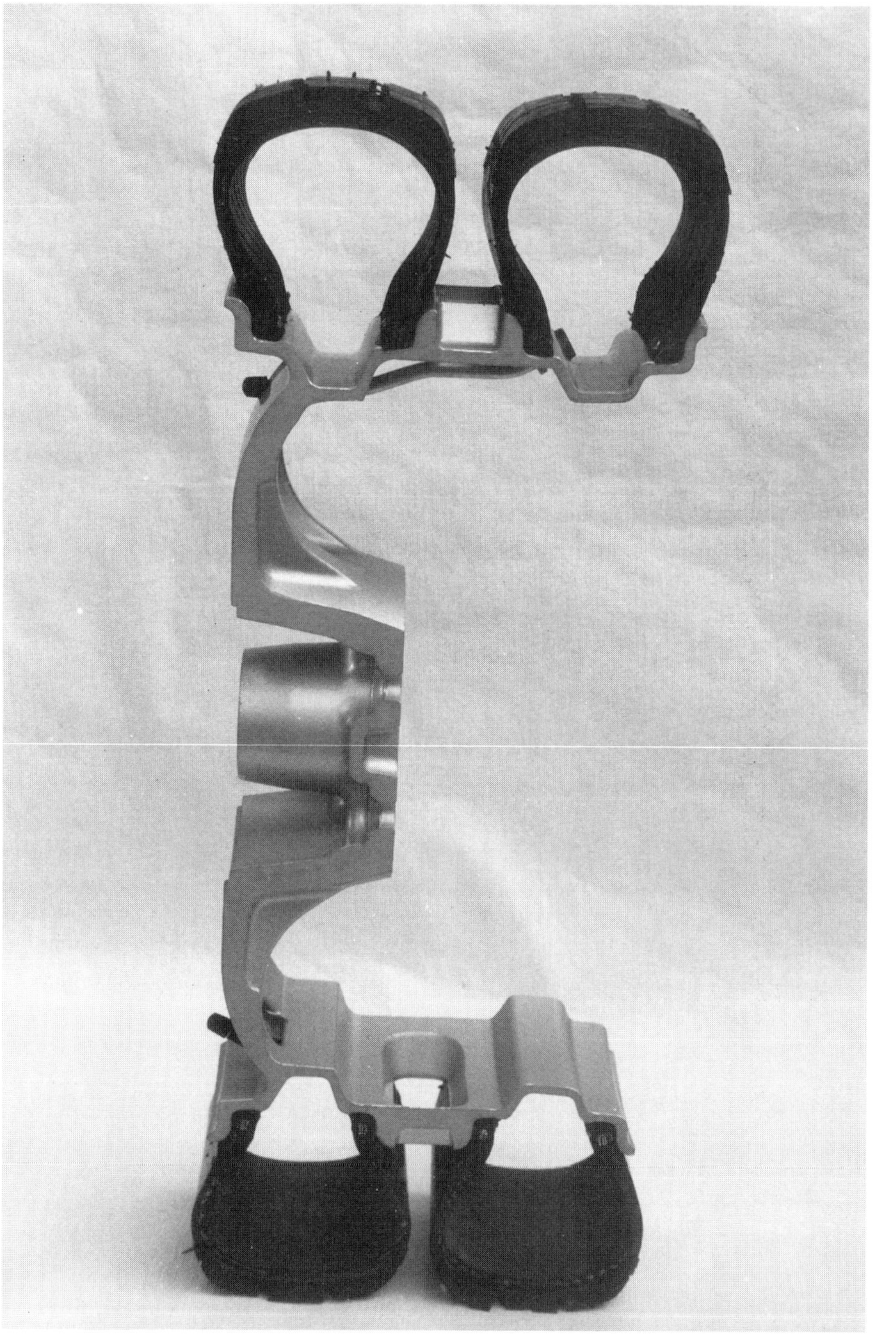

wheel to be dispensed with, without recourse to specific run-flat features (see later in this chapter).

In his first experiments, Juhan tried mounting two very narrow, steel-belted, radial tyres—125/85SR-15 as used on the Citroën 2CV—on specially-made wheels which he fitted to a Lotus Esprit. On its standard 235/60VR-15 tyres, the Lotus proved very prone to aquaplaning in heavy rain, but this problem disappeared with the twin tyres. The space between them acted as a gutter, and the narrow tyres themselves cut through the standing water more easily.

Encouraged by the success of these experiments, Juhan approached Goodyear's

International Technical Centre in Luxembourg who produced two sizes of narrow-section, V-rated (over 130 mph/210 km/h) tyres in sufficient numbers for serious development work to start. Speedline, a specialist wheel manufacturer in Padua, Italy, made a batch of alloy wheels on which the twin tyres were mounted.

As testing proceeded on Goodyear's track and on public roads, it soon became clear that the twin-tyre system had advantages other than the virtual elimination of aquaplaning. Tread life was surprisingly high, fuel economy was improved slightly, stopping distance decreased, and there were cornering benefits on wet and dry surfaces. In particular, a car shod with twin tyres seemed much less liable to change its attitude dramatically in power-on/power-off transient cornering manoeuvres.

The safety benefits were proved fully. If one tyre was deflated, stability was retained and the car could still be driven (at higher speeds and for longer distances than with competitive, run-flat systems) without too much effect on its handling, provided the remaining tyre on that rim was inflated to a much higher pressure to compensate for its higher loading. This was a point that worried Goodyear considerably. If the driver of a twin-tyred car had a puncture, but was unaware of the fact because he could detect no difference in handling, he might continue driving hard, oblivious to the fact that one tyre was well overloaded for its inflation pressure. Ultimately, that tyre would also fail, wiping out all the theoretical advantages. Goodyear's legal experts baulked at the product-liability implications if they were to become the only manufacturer of the JJD system and offered it without some form of built-in, foolproof, deflation warning.

Juhan dismissed the issue and terminated his association with Goodyear tyres. However, research into the twin-tyre concept continued, both at car makers who had become interested and by the German government-funded TÜV (Technischer Überwachungs-Verein Bayern) official testing institute in Münich. Avon Rubber Ltd, of Melksham, Wiltshire, the small, independent British tyre maker, began producing tyres for Juhan, who negotiated several other contracts in Europe, America and the Far and Middle East.

The police and ambulance services in Geneva were the first to adopt the JJD system, at the beginning of 1984, and have since professed themselves to be well satisfied with its performance. On at least one occasion, the JJD system has allowed a patrol car to continue chasing suspects after sustaining a puncture or burst in one tyre, and JJD tyres can often be spotted on government and security-force vehicles.

Although limited batches of experimental tyres had been made by Goodyear and Avon, the first production run of 1000 narrow, high-speed (125/90VR-16) tyres was made by Avon for JJD in 1985. These were tested extensively by Avon themselves on a Porsche 928, and by the TÜV; both reported impressive results.

While Speedline tooled up to produce a wider range of double-rimmed JJD wheels, Avon installed more equipment for producing narrow-section tyres to the JJD principle in a range of 16 in. (40 cm) rim sizes. Production of a 105/85VR-16 tyre began at the end of 1985, followed by a 115/85VR-16. With these sizes available, the range of cars to which JJD twin-tyre-system wheels could be fitted rose to almost a quarter of the European market.

Despite the safety advantages and added convenience of being able to drive on with a puncture, at the time of writing, no car maker had decided to offer JJD twin tyres as standard equipment, or even as a factory-fitted option. The problem for most mass producers is that they like to obtain their tyres from competitive suppliers to keep the original-equipment prices low. However, many JJD sets have been retro-fitted to cars in various European countries, especially in Germany where car makers have shown a particularly strong interest in the concept.

Negative factors of twin tyres, apart from their higher cost which can be offset against the deleted spare wheel, include a deterioration in ride comfort on some cars and heavier steering. The same problems tend to exist in cars using conventional, ultra-low-aspect-ratio tyres. Doubtless, they could be resolved by detail development and power-assisted steering.

The safety aspect remains the most controversial issue concerning twin tyres, and their more general acceptance will probably have to wait for the availability of a cheap, simple and reliable, radio- or infra-red-operated, deflation-warning system. The cost comparison between four twin-rimmed alloy wheels, with eight narrow tyres, and five conventional wheels and tyres should not pose a long-term problem.

Before the JJD system appeared, there had been several previous ideas on eliminating the dangers and inconvenience of a punctured tyre. Perhaps the most sensational concept of them all was the Dunlop Denovo, which was launched in the late 1960s as the first, real, run-flat design. It was intended to dispense with the spare and allow the driver to continue, at reduced speed, for up to 100 miles (160 km) with the carcass running against the rim. The system relied on a ring of small capsules of sealant/lubricant, carried inside the air chamber, being ruptured by the weight of the vehicle on the deflated tread, and positive bead location to prevent the tyre running off the rim.

Although Dunlop claimed that the necessary dealer infrastructure to support the Denovo would be available across Europe, and it was offered for a while as optional equipment on the BL Mini and the Rover SD1 range, it was highly priced and failed to make much impact.

In 1983, Continental Gummi-Werke released details of their revolutionary new tyre system, which has been under development ever since. Once the necessary licensing agreements are completed for it to be offered by several competing manufacturers, it looks set to succeed where the Dunlop Denovo failed.

The Conti Tyre System (CTS) concept wraps round the wheel rim to give a more evenly stressed carcass and special run-flat properties

CTS (Conti Tyre System) is a run-flat design which uses a sort of 'inside-out' wheel rim and bead: the tyre bead wraps around the outside of the wheel rim and is retained by a hollow-section insert. This has several advantages for the car designer and vehicle operator alike. It saves space in the spider/rim area of the wheel, where conventional rims have their well, allowing bigger brakes to be installed with more air around them. It takes the S-bend out of the carcass sidewall, allowing it to flex more easily for improved ride comfort (or lower sections to give the same ride comfort as before). According to the makers, it is ten per cent lighter as an assembly, reducing unsprung weight, has 15 per cent less rolling resistance, ten per cent more wet-road traction and 25 per cent less tendency to aquaplane.

Removing most of the rigidity in the bead area allows the footprint to develop more freely, with more uniform pressure distribution on the ground and more effective use of the tread. The energy wasted in the distortion of a conventional, stiff bead is saved, which aids fuel economy. But the real claim for CTS is its ability to stay wrapped round the rim when deflated. It is so firmly locked on the rim, say Continental, that it is possible to maintain full control and drive for several hundred miles on a flat CTS tyre, eliminating the need for a spare.

The arguments for and against spare wheels have been raging for many years, and all were voiced when the Denovo was launched. When most drivers get a puncture, they want to fit the spare as quickly and easily as possible, and get motoring again without restrictions. The space-saver wheel and tyre, supplied with Audis, Porsches, Golf GTis and Saabs, have conditioned quite a large section of the motoring public against that ideal. Furthermore, tyres are so resistant to punctures now that the statistical average incidence is only one puncture in every five years.

The CTS, it seems, can really keep the wheels turning as far as the nearest tyre dealer. That was the idea Dunlop proposed, but there is a key difference in the way Continental are handling their patent. Instead of keeping it to themselves to give them an edge over their competitors, they are selling licences in a big way and working very closely with all the major car producers.

CTS tyres and wheels can be driven flat on their rims safely for several hundred miles at moderate speeds. Eventually such designs could eliminate the universal but mostly redundant spare wheel and tyre

By the end of 1985, Continental had signed up enough tyre companies to cover up to 65 per cent of the total European market for original-equipment and replacement tyres, going far beyond their own brands (Conti and Uniroyal) and Semperit (who joined forces with them early in the project). They have also spent time selling CTS to the vehicle manufacturers, continuing development at full speed with the help of the Porsche R & D Centre at Weissach, the Technical Control Board in the Rhineland and two technical institutes in Hanover University. They even gained a subsidy from the German Federal Minister for Research and Technology. Both Ford and Volkswagen are developing new vehicles using CTS.

By the time it appears on a production car in 1988, CTS will have cost about £23 million ($35 million) to develop, but Continental believe the investment will pay ample dividends in the future. The development time from announcement to production will be seven years, but they expect CTS to take over completely from conventional tyres by the mid-1990s—about a third of the time it took radials to take the place of cross-plies.

8 Racing-car suspension systems

In the very early days of motoring, it was the competitive spirit of drivers who raced their cars on the public roads that provided the main stimulus for suspension development. Speeds were limited at the time, as much by the ride of the primitive racing machines on the appalling surfaces as by the power of the engines and the brick-like aerodynamics of the square-edged, upright bodywork.

As the heroic drivers wrestled with huge steering wheels, skidding on loose gravel and bouncing from crag to ditch, suspension design progressed in similar leaps and bounds. Over a very short period, with little application of kinematics and varying standards of execution, springs developed from the long, curvacious, double-elliptical blades of the horse-drawn carriage to the more substantial, multi-leaf arrangement of semi-elliptics, three-quarter-elliptics, short cantilever elliptics and every other permutation in the book.

During this period, car designers had little choice in the materials and manufacturing methods available. Springs had to be tough to withstand bumps and potholes, and there was little need for more positive, axle-locating links. Stage by stage, design polarized towards the logical and effective semi-elliptic leaf, usually installed longitudinally on each side of the car, but sometimes mounted transversely, particularly at the front. As mechanical trail angles (castor) brought improved stability, steering efforts increased and tiller systems were abandoned in favour of the more manageable wheel.

Of course, there were notable exceptions. The first independent front suspension, for example, appeared in 1898 on a French Deauville lightweight racing carriage (now in the Museo dell'Automobile, Turin) in what became known as the *voiturette* class. It had sliding pillars on a cross-tube support with a slim, transverse, leaf spring mounted above. The design was copied in 1902 by the pioneering French family of Amédée Bollée et fils (they had started with steam carriages as early as 1873) who added it to a sports car built for the 1899 Tour de France by Clément Ader (the telephone engineer). It was also imitated by Sizaire-Naudin in the touring car which ran in the 1906 Coupe de l'Auto *voiturette* races.

Another exception to the general rule of cart axles mounted on leaf springs appeared on a range of popular runabouts built by Count Albert de Dion and Georges Bouton, and introduced in 1894. They used a system which transmitted the drive via two universally-jointed half shafts mounted independently of a lightweight, tubular axle attached to the wheels. It appears to have been copied from an early steam carriage designed by Amédée Bollée and gave the same advantage of reduced unsprung weight as the chain-driven systems used extensively

Semi-elliptic springs and the simplest design of shock absorber proved fairly adequate in the early days of racing. The introduction of front-wheel brakes added considerably to the unsprung masses and caused a host of new systems

Frazer-Nash used a sliding trunnion at the rear of their semi-elliptic system on their 1930s racing Chain Gang. The short, forward-facing link parallel to the leaf spring is an attempt to control axle wind-up under braking

on large cars. Some 40 years later, the de Dion axle was destined for a significant revival.

Another development, introduced long before its value could be appreciated fully, appeared on a racing Panhard in 1902. It took the form of a transverse radius rod used to locate the front axle and allow a softer transverse spring to save weight.

The Grand Prix races organized by the Automobile Club de France, which started with the 1895 Paris–Bordeaux–Paris road race, became the breeding ground for racing-car design. Within a few years, engines had capacities of 11 litres or more and were installed in lightweight chassis, which were still made with wooden side rails reinforced with steel plates, weighing less than 2200 lb (1000 kg) in total.

Front-wheel brakes had not been deemed a necessary complication yet (braking on the rear wheels and transmission were considered sufficient), so unsprung mass at the front was kept to a minimum. At the rear, chain drive to each rear wheel provided the only reliable transmission for high-powered cars, and brought many other benefits. It allowed these monster machines to use relatively-light, tubular rear axles on which were mounted free-running wheels. By using a chassis-mounted final drive, differential and rear braking system, unsprung masses at the rear were also low, resulting in unexpectedly good roadholding and handling.

Friction dampers, as a supplementary part of the total suspension system (leaf springs operate with inherent inter-leaf friction, giving a slight degree of integral damping), first appeared on a Brasier-designed, French-built Mors racing car in

1903. Driven by Gabriel, it ran in the ill-fated Paris–Madrid race.

Before leaving this formative era of early racing-car design, Bugatti's characteristic, reversed, quarter-elliptic rear springs should not go without mention. This layout was introduced as early as 1911 and persisted on his cars as late as 1950. Although no official or deduced explanation has ever been suggested for this arrangement, its installation on a rear-biased, narrow cross-member would seem to generate deliberate frame flexing and reduced rear roll stiffness (similar characteristics, in fact, to the much more modern, trailing twist axle so common to modern front-wheel-drive designs). On a Bugatti, the effect would be to counter oversteer caused by the combination of high power and rear fuel-tank mass.

In the 1920s, emphasis centred on engine design, most designers staying with the time-honoured and predictable combination of ladder-type chassis frame with semi-elliptic springs all round. One notable exception was the 1922 Benz RH derived from Edmund Rumpler's Tropfenwagen touring car.

The RH (Rennwagen Heckmotor) was a six-cylinder, mid-engined racing car with swing-axle rear suspension and very advanced, aerodynamic bodywork designed by Hans Nibel and Max Wagner. It had some successes in Grand Prix racing before its development was cut short in 1926 by the merger between Benz and Daimler, the latter preferring to develop front-engined racing cars designed by Ferdinand Porsche. It is ironic that, when Porsche left Daimler-Benz to form his own design organization and was busy developing the mid-engined Auto Union, Nibel and Wagner became responsible for the front-engined Mercedes-Benz Silver Arrows.

Start of the modern period

The classis period of Bugatti dominance in Grand Prix racing coincided with a time during which the other works teams had lost interest and retired from the fray. The late 1920s and early 1930s saw a period of stagnation in European Grand Prix car design, which was rudely dispelled with the advent of the Nazi-sponsored Mercedes-Benz and Auto Union teams. German engineers were only too glad of any opportunity to exercise their skills, and with ample funds available they created many advanced chassis concepts, inspired by newly-developed, passenger-car layouts.

In designing the suspension for these cars, the aim of both companies was to keep overall vehicle weight as low as possible. To this end, Mercedes-Benz, who had already produced successful passenger cars with the same layout, opted for a front-engine, swing-axle rear suspension and parallel wishbones at the front. They achieved a low driver position by giving the car, the type W25, a chassis-mounted, rear transaxle (combined gearbox and final-drive unit) with the propeller shaft offset downwards and passing close below the driver's seat.

At the front, the W25 had a beautifully-constructed wishbone system with coil springs located inside a transverse cross-member, which was machined from a solid steel billet. Independent front suspension allowed the engine to be installed well forward in the frame to give the car stability and keep the overall length down.

The Mercedes-Benz suspension was characterized by extremely-limited wheel travel, the total movement being only 1.75 in. (4.5 cm) at the front and 2.75 in. (7 cm) at the rear. Such limitations were typical of leaf-sprung racing cars at that time and were largely due to a cautious approach to undue geometric freedom because of the wheel camber changes inherent in swing-axle design. Even with this

4·5

Matra used their aviation stress engineers to design a complex system of triangulated links for the rear suspension of their 1970s Grand Prix car. Cast-magnesium hub carriers were reinforced by external tie bars to protect against material failure

limited movement, which called for high-rate springs that nullified any of the advantages independent suspension might have imparted to the ride, the very high, rear roll centre, located some 2 in. (5 cm) above the axle centreline, caused high roll moments and a pronounced tendency towards oversteer.

The contemporary and competitive Auto Union consortium sought a more advanced solution for their Porsche-designed car. It drew its inspiration from the Benz RH, achieving a low driver position and minimum frontal area by putting the engine behind the driver and mounting the transaxle unit directly on the rear of the engine. Like the Mercedes, it used leaf-sprung, swing-axle rear suspension, but this was combined with double-trailing-link geometry at the front.

There were very similar oversteering problems on the Auto Union as well, exacerbated by the forward location of the driver who had little sensation of what was happening to the back end of the car until tail swing had progressed too far. While a long-term engineering solution was developed, the immediate problem was solved by seeking out a driver who had no preconceived ideas about how a racing car should react. They found him in ex-motorcyclist Bernd Rosemeyer, a sensitive and courageous driver who had a natural ability to use the full performance of the arguably tricky cars.

The solution to the oversteer problem for both teams was to adopt a modernized version of the de Dion axle, which lowered the rear roll centre and kept the rear wheels parallel at the same time, relieving them of geometric camber changes. It was also possible to increase wheel travel without running into handling problems, so both teams began to use lower-rate springs. On the Mercedes-Benz W125 and W163, Nibel and Wagner virtually doubled the suspension movement of the W25 by redesigning the front suspension with longer wishbones and external coil springs. The Auto Union's handling was much improved by relocating the driver further back in the car.

Mercedes achieved lateral location of the de Dion axle beam with a sliding trunnion mounted on the rear of the gearbox casing. Auto Union chose a Panhard rod with its anchorage set low on an extension of the chassis frame, positioning the rear roll centre within a few inches of the ground. Both teams used deep-section, trailing links to provide fore-and-aft location of the de Dion beams and absorb braking torque, but Mercedes made no attempt to reduce the rear roll height. Presumably, they adopted the de Dion layout solely to keep the rear wheels upright without the characteristic jacking effect so typical of swing-axle design.

Both cars displayed consistent understeering behaviour once their swing axles were abolished. A major gain for Auto Union was that this gave them a wider choice of drivers, which was particularly fortunate since Rosemeyer was killed in a rather futile record attempt late in 1937.

During this period between the wars so much new ground was broken in engineering, and such was the impact of German technology on racing-car design, that a pattern was set which persisted for almost 20 years. After a poor start, Mercedes-Benz proved the benefits of the relatively-soft, long-movement suspension with their second-phase cars. In turn, Auto Union confirmed the significance of the mid-engined Benz RH as the forerunner of the modern Grand Prix car, and with their low-mounted Panhard rod emphasized the importance of a low, rear roll centre for this configuration. The mystique of the Stuttgart and Zwickau racing cars continued to influence design concepts for over a decade.

Post-war developments

In a strange, unrelated way, the war effort in Britain had a significant effect on post-war design. Apart from rapid developments in supercharging and turbocharging for aircraft engines, the British campaign led to a plethora of flat, rural airfields with concrete or tarmac runways and winding perimeter tracks. When peace was declared, the country suddenly found itself with dozens of ready-made motor-racing circuits.

The opportunity for the British to go motor racing on roads with hard, flat surfaces and challenging corners was something that had been denied them by law since motoring began. Before the war, there had been Brooklands with its banked turns, the Shelsley Walsh hillclimb and the narrow roads of Donington Park, but real road racing had been confined to Ireland, the Isle of Man and continental Europe.

The new wave of British enthusiasm for motor sport rode in on the backs of Silverstone, Goodwood, Snetterton and many less-famous venues scattered around the country and used as training grounds for drivers and designers. By fair means or foul, European sports and racing cars were imported to Britain, hundreds of new car producers sprang up overnight, and historic names turned their attention to the new craze. Along with all the private and sponsored efforts, the Owen Organization created a new team called BRM, financed by public subscription.

An important move in the motor-racing story was made by the Royal Automobile Club who introduced a new low-cost, 500 cc formula, based on motorcycle engines. By attracting designers, constructors and drivers from all levels of society, this was probably one of the most significant developments that ever happened in motor racing. Inevitably, the best came to the top: the Coopers as constructors; Stirling Moss, Mike Hawthorn and Peter Collins as drivers; Norton as the predominant engine. But the real gain derived from 500 cc racing was a new approach to racing-car design, based on light weight, small bulk and the science of advanced suspension design.

Cooper pioneered the system of inboard front springs operated by lever arms on their 1966 Maserati-powered Formula 1 design

Denied the typically-German, high-technology approach to racing-car design, and access to elaborately-equipped, precision-tool rooms staffed with engineering craftsmen, these newcomers (many of whom had spent the war in aircraft design offices) soon learnt new, more scientific ways to make their racing cars corner quicker than their rivals. They also discovered that readily-available materials and ordinary workshop techniques, properly applied, could give results equal to those from the factory racing departments.

The conceptual lessons learnt in 500 cc club-class racing took time to reach the Grand Prix scene, which was steeped in complex, front-mounted engines with up to 16 cylinders and two-stage supercharging. While the smaller mass and bulk of the midget single-seater was an excellent tooth-cutting exercise, and a splendid spectacle for the enthusiastic crowds, suspension design on these cars stabilized quite early to the Cooper system of simple, independent, transverse leaves.

The real progress in racing-car suspension came from sports-car racing in the 1100 cc and 1500 cc classes contested by Lotus, Lola and Cooper. By clever use of space-frame structures and aerodynamic, all-enveloping bodies, these companies produced lightweight cars capable of very high speeds—and the problems of getting them round corners as quickly as possible began to exercise a number of fertile, uncluttered minds.

By the end of 1968 the Cooper lever-arm system had been refined and much copied in racing circles

Rear-suspension systems

Suspension systems for the rear wheels of a rear-wheel-drive car have always been compounded by the need to provide a mechanical driving system. In the main, the second phase of suspension development, in the late 1950s, concentrated on this end of the car. As early as 1930, Daimler-Benz had realized the importance of keeping the rear wheels of a racing car parallel to each other and at right angles to the ground. The young British designer/drivers took this principle one step further by designing linkages which, with the round-section tyres of the period, could provide a little roll-induced negative camber on the outside wheel to increase the cornering power of the tyres.

To promote negative camber angles with roll, Colin Chapman (the late founder of Lotus Cars) introduced an original, strut-type rear suspension based on established MacPherson principles. Careful selection of the location of the inboard pivots for the lower links provided a camber change of about 2 degrees. This became known as the Chapman strut and was applied successfully to the World Championship-winning Formula 1 Vanwall in 1957 and 1958, and to the original 1957 Lotus Elite coupé. The space occupied by the inclined springs became more important than the geometric advantages when centrally-mounted engines and wider tyres showed their superior track performance.

The next important development came from the drawing board of Eric Broadley, a new designer whose 1100 cc Lola sports racing car had proved the most successful in 1956 track events. A key factor in the Lola design was its revolutionary, low-pivot rear suspension in which the fixed-length drive shaft acted as the upper link of what was, in effect, an unequal-wishbone layout. The triangulated lower link of this system was set very close to the ground to achieve a reasonable vertical displacement between the two geometric links, lowering the rear roll height at the same time. A single radius arm on each side, pivoting on the same centres as the propeller shaft, acted as the third side of the upper triangle and transferred fore-and-aft loads to a point well forward in the frame.

The principle of spreading suspension loads over a large area of the chassis was a significant step forward which has been followed ever since. Before, racing-car suspension had followed road-car practice more closely, input loads being transmitted to the frame over short distances in a relatively-small, heavily-reinforced area adjacent to the wheels. Broadley saved much of this weight and gave his long, lower radius arms a considerable mechanical advantage by locating their pivot points well ahead of the rear wheels.

It still needed the Cooper-Climax racing cars of the late 1950s to break the spell and prove that a small, light and simple mid-engined car was more than a match for the complex creations of the establishment in racing-car design. By placing the driver ahead of the engine, as Porsche had done with the relatively-complex Auto Union, Cooper made massive improvements to the frontal air flow, easing penetration and reducing the drag coefficient significantly. Further refinements were added by particularly neat, wishbone suspension units.

Another lesson taught by Cooper was the advantage of being able to set the suspension to suit each specific circuit. This was achieved by providing alternative suspension pick-up points and adjustable links, so that each part of the controlled wheel movement could be varied. Of course, it took a skilled test engineer/driver to optimize the settings and led directly to a new breed of racetrack hero.

As the Cooper concept graduated to larger cars, which began to win major races

This Lotus Formula 1 design from the 1970s shows how the engine/transmission had become a structural part of the car, supporting an oval suspension subframe and the lower wishbone attachments

in many different categories, mid-engine locations became the norm in time for a new 1.5-litre Grand Prix formula. At the same time, the Broadley suspension layout that appeared on the Lola 1100 came into its own.

From this point, racing-car design began to polarize again, with variations in detail rather than basic concept. Colin Chapman showed his innovative flair on the 1960 Formula 1 Lotus, where the lower wishbones were reversed and a second, lower radius arm added to match the controlling influence of the drive shaft, which acted as one link in the upper triangle. This elegant solution allowed the lower, reversed wishbone, which was provided with adjusters and was attached at two well-spaced points on the bottom edge of the cast-aluminium wheel carrier, to locate the path of the rear wheels precisely, leaving the trailing radius arms to resist the braking and torque loads.

Chapman's masterly design was fundamental and showed the way forward. It provided a simple, trackside method of adjusting the characteristics of the rear suspension as well as accurate rear-wheel location with the minimum unsprung mass. The same kind of long, trailing radius arms, mounted well forward, were destined to become even more useful when chassis designers turned to the double-sponson type of monocoque construction where the engine forms the main rear structural member.

When American tyre manufacturers entered the European motor-racing scene in the late 1960s, European racing-car designers were faced with another new problem. The new, ultra-low-profile tyre sections and rubber compounds created for the high speeds of Indianapolis and drag racing made it even more important to keep the tread footprint flat on the road. Some of these tyres, designed for dry weather only, were totally devoid of the normal tread pattern, so it was of paramount importance that the full width of the tyre was in contact with the road under all conditions of bump, rebound and roll.

The change required careful reassessment of the geometry, causing the length of the upper wishbones to be reduced to almost half the length of the lower ones at both the front and rear of the car. This pulled the top of the wheel inward under bump movement to counter the natural body roll. For this reason, the idea of using fixed-length, rear drive shafts as suspension links had to be abandoned in favour of variable-length shafts able to 'move' in and out by means of either sliding spline joints or a flexible rubber 'doughnut' at the inboard end. To fine-tune these characteristics, the addition of an adjustable upper wishbone became imperative.

There have been innumerable variations of what became known as four-link rear suspension. Quite early on, it was realized that a reversed upper wishbone plus a short, adjustable tie rod—forming an N configuration—was a convenient way of providing adjustable, rear-wheel toe angles. Usually, this system was associated with a variable-length, single-tube lower link, which was used to set static wheel camber. An alternative arrangement, introduced by Cooper, had a variable-length, upper radius arm and a Chapman-type, reversed lower wishbone adjusted by screw-in Rose joints at the hub-carrier ends. Most designers had their own variation, but the principle of providing adjustable rear suspension with widely-spaced load inputs remains the same to this day.

This early Ferrari design is another typical example of a four-link rear suspension. It used long forward-facing radius arms combined with unequal-length top and bottom transverse links to form a classical wishbone layout

By 1986, aerodynamic development for Formula 1 had slimmed down the nose considerably, demanding long, slim suspension arms that would not obstruct air flow to the pannier radiators and turbo intercoolers

Front suspension

There have been fewer variations at the front of the car. Trailing-link front suspension disappeared with the Auto Union and double-wishbone layouts became virtually universal. When coil springs took over from leaf springs, they tended to be exposed to the airstream, a practice which persisted until the late 1960s. As engine power rose, the effects of turbulence between the open wheels and pencil-shaped bodywork was not considered a major issue worth the trouble of eliminating.

However, when monocoque structures replaced tubular space-frames, the car's total frontal area was reduced almost to that of a recumbent driver, and even small reductions in drag were seen to make a difference. At this stage, the universal coil-spring-and-damper units were moved inside the nose section and actuated by levers formed from extensions of the upper suspension links. During this period, a common practice was to weld a sheet-metal skin across both sides of the wishbone tubes, boxing the space between them to increase their stiffness and reduce drag still further.

Modern practice in the late 1980s

During most of this decade, there has been a steady trend towards cars with side radiators, narrow needle noses and excessively wide track widths. These features require extremely long wishbones to separate the wheels from the body by the maximum distance for a clean airflow to the various intakes and aerodynamic devices. Springs are housed inside the nose and often mounted horizontally across the car, being operated by lever-arm extensions of the upper wishbone links or pull-rods acting on the fulcrum of a separate pivoting beam.

Dampers are adjustable for their internal effects and their stroke, which sets the static ride height of the car. Either the twin-tube or single-tube systems are used, the current trend being towards the superior performance of the latter type (see Chapter 4). Even on today's smooth track surfaces, dampers still seem to have heat-dissipation problems.

Aerodynamic requirements of the modern Grand Prix body also call for a free flow of air around the rear of the car, so the most usual practice is to mount the rear springs vertically and close to the sides of the transmission. They are actuated by brackets located part way along the lower suspension links, which pivot close to the car's centreline.

Structurally, these wishbones are a far cry from the little, tubular triangles used in the immediate post-war period. The lower ones are in the form of large, open, equilateral triangles with sides measuring about 3 ft (1 m) long. Construction of the upper wishbones depends on the way the springs are actuated. If the wishbone is extended to act as a lever, its beam strength has to be increased by making it with a deeper section. Otherwise, it is fabricated from round-section tubes. It is interesting that no attempt appears to have been made yet to utilize the smooth, oval-section tube which was once available to the light-aircraft industry.

The considerations influencing the geometry of racing-car suspension have changed very little during this decade. The main requirement is still to keep the full width of the tyres flat on the road under all dynamic conditions. The second criterion is to maintain a reasonably-constant ground clearance against the aerodynamic downforces, which can be as much as three times the static weight of the fully-laden car.

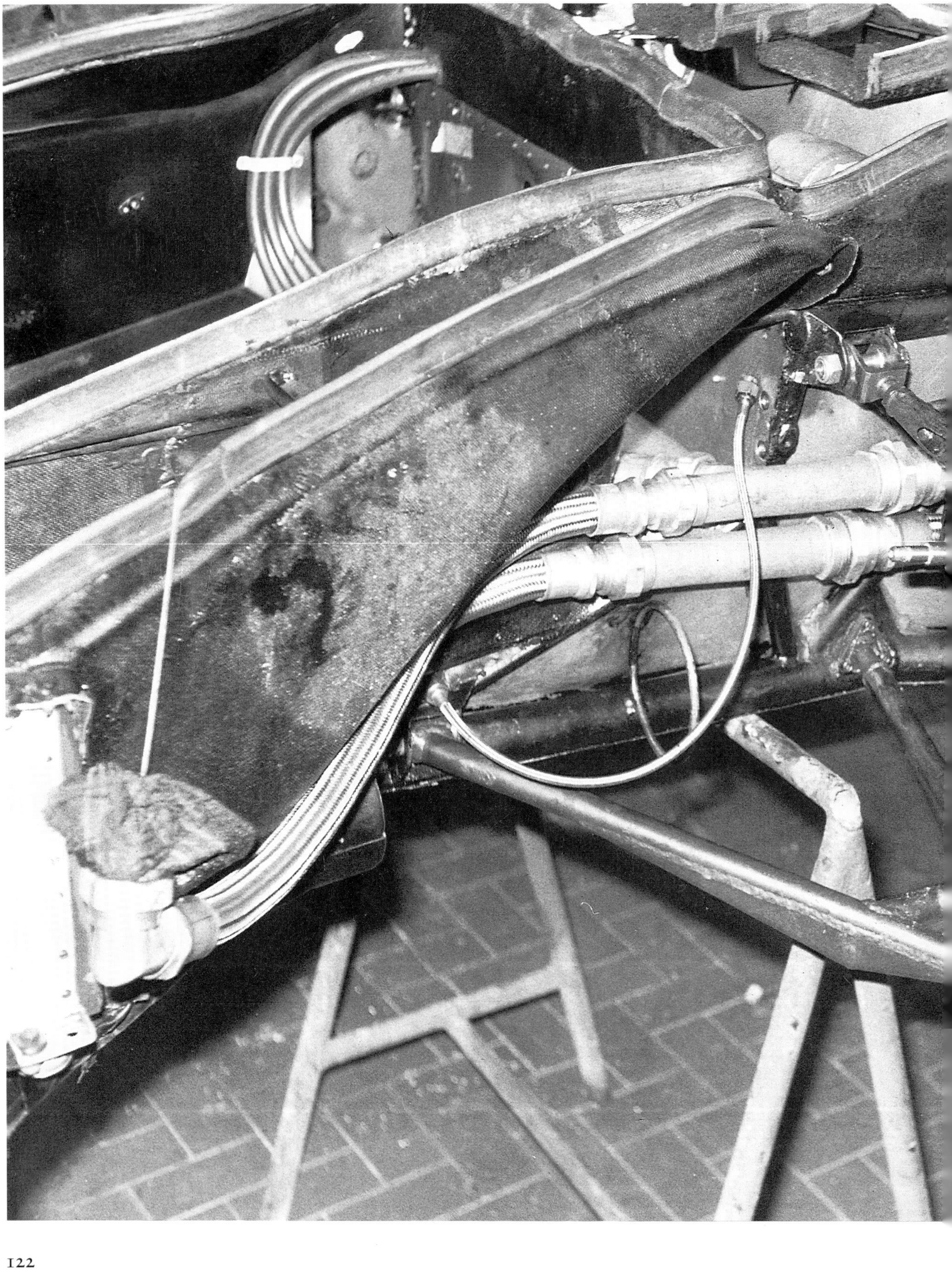

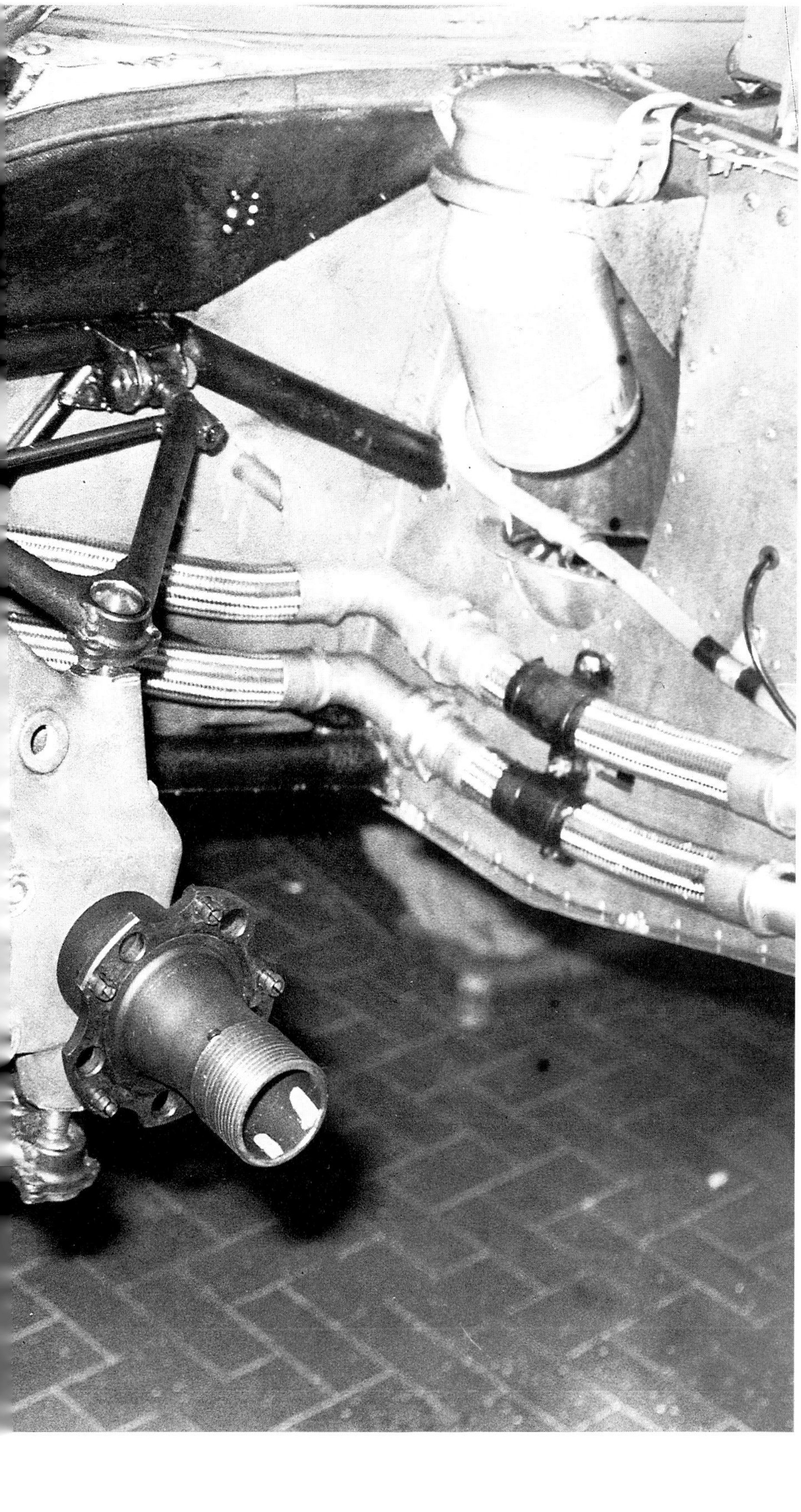

On the sports-racing Ferraris of the 1970s a double-wishbone system provides short upper arms and long lower ones, to keep the wheel and tyre upright when the body rolls

This early Brabham Formula 1 design eliminates buckling loads from the upper and lower radius arms under braking. The long upper arm faces rearwards and the lower one forwards so that both links withstand only tension

Some success in maintaining clearances is achieved by raising spring rates as much as possible without increasing ride frequency beyond about 1.3 Hz. The problem with unequal-length wishbones is the shift in roll height between full-bump and rebound positions. Usually, the smooth track surfaces available and restriction of total suspension travel to around 3.25 in. (8 cm) keep this problem under control and inhibit camber change at the same time. It is only on some of the relatively-rough road circuits, like Monaco and Detroit, that suspension weaknesses are revealed.

Before leaving the subject of modern racing-car systems, mention must be made of the electronically-controlled, hydraulic suspension developed by Lotus in conjunction with the Cranfield Institute of Technology. This research, initiated in an attempt to overcome the suspension problems created by the huge variations in aerodynamic loads, is now finding a new significance for production sports cars (see Chapter 9), particularly after the acquisition of Lotus Cars by General Motors in 1986.

When the system was first conceived in 1982, aerodynamic downforces at maximum speed amounted to about two-and-a-half times the static laden mass. In cornering conditions, lateral accelerations of 3.5–4.0 g were also possible with full ground-effect skirts. To withstand this, and to minimize aerodynamic and geometric changes created by the natural pitching and rolling motion, exceedingly-high spring rates and very hard damper settings were required. Although the resultant cars were very responsive, they were uncomfortable to drive, and the vibration combined with high g-forces frequently caused minor spinal and neck injuries to the drivers.

The object of the Lotus/Cranfield active-ride system was to overcome these problems by replacing the road springs with heavy-duty, double-acting hydraulic actuators controlled by a microprocessor to which was fed a series of dynamic inputs. Hydraulic pressure was provided by a gearbox-driven pump, and the height of the car controlled by a linear, variable-displacement, electro-hydraulic servo valve and a system of load cells developed for this particular application by the Cranfield Institute.

In action, signals from the load cells were processed by the computer to simulate the effects of damping and springing combined and fed to servo valves in the actuators, which could operate over a wide range of internal pressures. In contrast to previous active-ride systems, which in general could only respond in time to cope with roll and pitch variations, the Lotus suspension used an advanced 32-bit computer to control the ride over surface variations and generate negative roll behaviour to increase stability and ultimate cornering power. Under heavy braking or fierce acceleration, the system could also anticipate and counter dive and squat effects caused by dynamic weight transfer.

The system was used experimentally in 1983 in the Brazilian and United States Grands Prix, before continuing as a factory development progamme with considerable future potential. It was not used again in competition until 1987 for fear that, as with so many exclusive and successful devices before it, pressure from other competitors would cause it to be banned. At that stage, it had also suffered from excessive weight and the need for computer processing power which had yet to be developed in a commercially-viable form. However, its potential has been proved so thoroughly that it is now a firm element in Lotus' future plans.

General design procedures

Although the early racing-car constructors used little more in the way of technical aids than a few chalk lines on the workshop floor, today's racing engineers must be fully versed in the high-tech world of computer simulations and advanced design aids. Not only are portable micros a familiar part of every Grand Prix team's pit equipment, but most drivers rely totally on full, two-way radio contact throughout each event. Load and temperature sensors in the car transmit readings telemetrically on all operational factors so that they can be monitored by the pit crew.

In terms of basic design procedures, the layout for a suspension system usually starts with some simple calculations over a range of cornering forces and the overturning couples. This envelope of data defines the transverse elevation of each suspension system and determines the highest safe point for the roll-height position. In other aspects of roll-axis angle and wheel-geometry design, conventional rules, as outlined in earlier chapters, are followed closely.

As with so much of modern car design, experience with previous systems always forms the basis on which further development depends. Even the best scientific theories are seldom found to function precisely in practice. Therefore, most of the racing-car world is dominated by a handful of highly-skilled and widely-experienced designers who have proved their competence and worth over many seasons' successes. For valid commercial reasons, they keep their secrets to themselves, command very high financial rewards and somehow seem to stay undisturbed at the top of their profession.

9 Active and adaptive ride

It has always been extremely difficult for the suspension engineer to maintain simultaneously a high standard of ride, handling and body attitude control under all driving conditions. The problems stem from the wide operating range created by the many possible combinations of road surface, speed and vehicle load. As the relative unsprung masses increase with reduced vehicle size, the problem becomes increasingly acute on smaller cars.

Conventional suspension systems must always be designed as a compromise. Even with the aid of self-levelling systems and interconnections between the wheels, any system that is soft enough to provide a comfortable ride cannot also provide a sufficiently-stable location of the vehicle body to ensure the best possible handling. Right at the heart of the issue is the often-overlooked and fundamental fact that ride is a measure of the ability of the suspension to handle mostly vertical forces applied at ground level, while handling and attitude control are influenced much more by horizontal forces acting on the centre of gravity and by ground-level couples or moments.

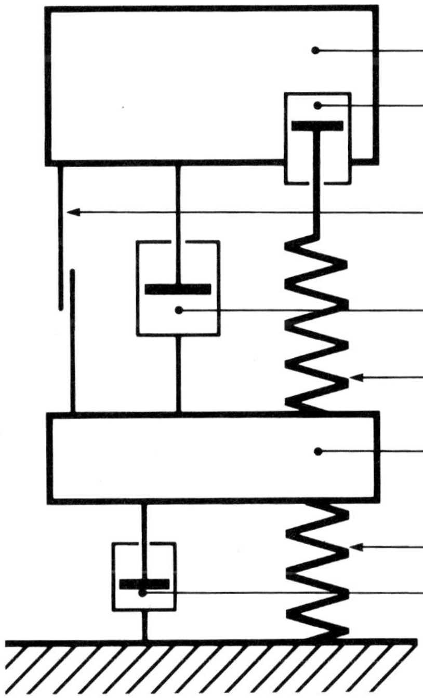

The basic elements of an active-ride system must include a quick-acting means of adjusting the ride height. The hydraulic strut above the upper spring mount performs that function in the schematic diagram shown here

The dynamic performance of suspension systems has improved progressively over the years, just as the science of wheel control has developed hand in glove with more-advanced analysis techniques and refinement of basic kinematics. The multi-faceted design of wheel geometry, springing media, locating members, damping elements, insulating bushes and tyre characteristics can achieve standards today that are way beyond even those envisaged through complicated hydraulic systems 10 or 15 years ago.

The fundamental difference between an active and a passive suspension is that an active system treats the road-induced forces separately from the body-inertia forces. By separating the functional elements of the suspension's task, active systems have the ability to eliminate the traditional compromise between ride and handling characteristics completely.

Strictly speaking, an active suspension is reckoned to be one which uses a fast-acting, closed-loop control system, such as those employed for engine management on today's high-tech engines. However, an active suspension can also be defined more simply as a system which is able to change its characteristics under outside influences to enhance its dynamic performance. In the modern sense, active or adaptive suspensions are driven by sensors and control systems which react, process and then operate a variety of fast-acting devices.

In every case, the objective is to reduce or, if possible, eliminate some of the compromises forced on the designer by a conventional passive system. In general, these relate to the limitations imposed on the designer by load, wheel travel and relatively-constant spring rates. In dynamic terms, the most restricting of all is the fixed relationship between pitch and bounce frequencies. A low bounce frequency for maximum ride comfort always leads to a disturbingly-low pitch frequency and usually excessive body roll.

As well as being system-driven, it can be argued that some of the less-active suspensions can also be road-driven, and there are several examples where the performance of active systems is simulated quite well by internal processing of the direct road inputs. The two examples which come to mind immediately are the famous Citroën 2CV, with its interconnected mechanical springs, and the ill-fated Morris 1100, which was the first car to use Alex Moulton's Hydrolastic suspension.

Within this definition of road-driven systems, Hydrolastic was one of the neatest and most under-exploited active suspensions ever conceived. It was patented by Alex Moulton in 1955 and launched commercially in 1962 as a key feature of the highly-acclaimed and innovative, front-wheel-drive Morris 1100. Although it worked very well in its original form, in service it suffered from long-term durability problems and never made a successful transition to the smaller Mini and larger 1800 hatchback and 3-litre saloon models. Later, it was developed into the Hydragas system, using the same principles but without the rubber springs, and appeared on the 1973 Austin Allegro. By rights, it should have become the mainstay of the BL technology revival, but its destiny was in the hands of the wrong people. In the Mini Metro, introduced in 1980, the Hydragas units were separated from each other to act as individual springs.

Dynamically, Hydrolastic was road-driven by the load inputs from the wheels and controlled totally from within the system itself. It had some very clever features that effectively simulated an active system, developed by a unique co-operation between Dunlop (who were responsible for the rubber cone springs), Alex Moulton (who conceived the original design) and Alec Issigonis (who fathered the complete project).

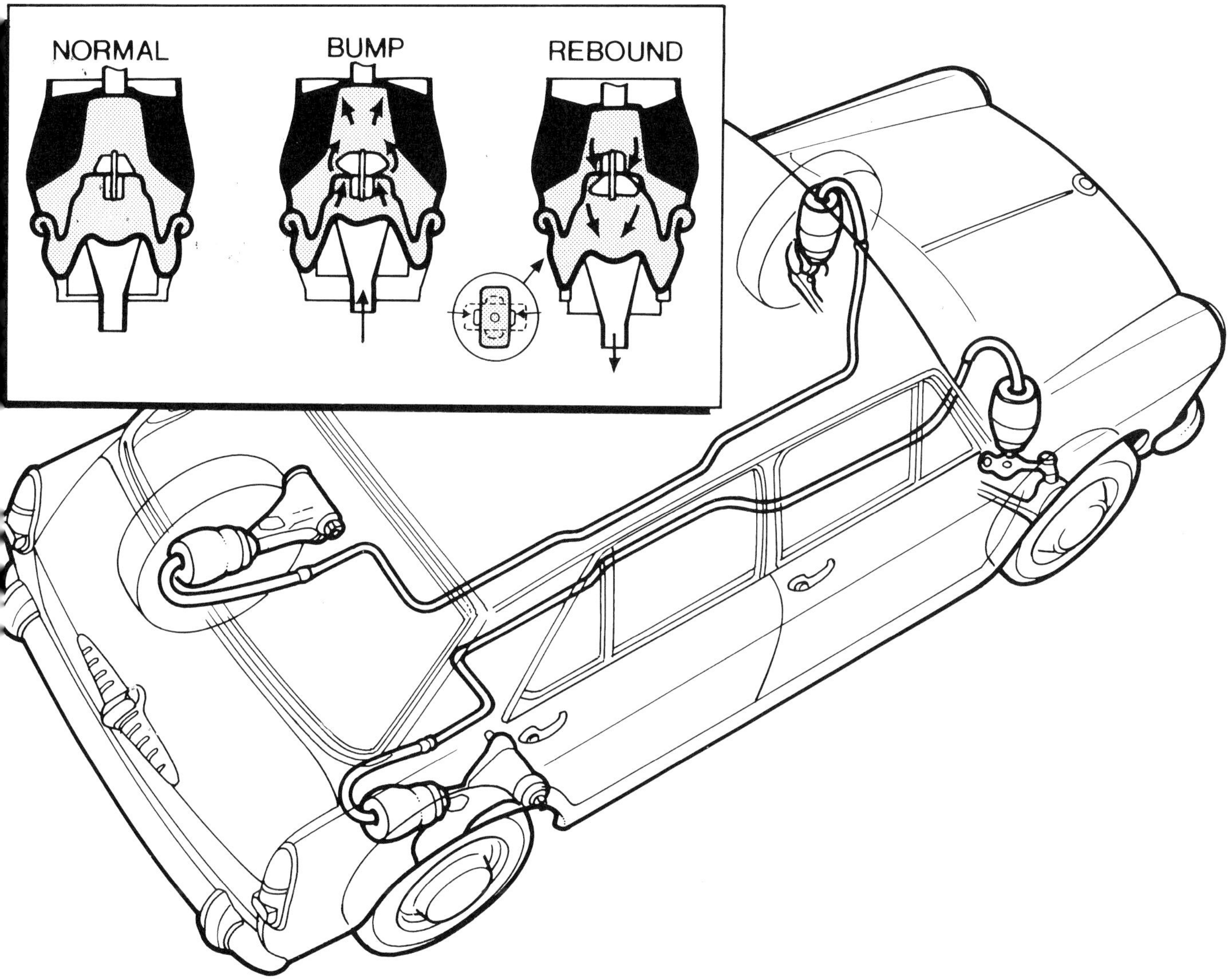

In geometric terms, the Morris 1100 suspension was pretty well conventional: double, unequal-length, front wishbones and simple, fabricated, rear trailing arms. As on the Mini, the rear springs were near horizontal and operated by lever arms.

Orthodox steel springs and hydraulic dampers were replaced, however, by the integrated Hydrolastic system, which carried out four specific operations: it absorbed vertical shock forces, it damped the subsequent rebound motion without fade (being water-based, the working fluid had a constant viscosity), it dissipated the energy of the induced motion, and it fed front inputs to the rear in proportion to the vertical wheel velocity. The result was a system that provided both variable-rate springs and variable-pitch reaction. Pitching moments depend upon pitch stiffness and the wheelbase, which sets the leverage effect of front and rear inputs about the centre of gravity. A short-wheelbase car is very stiff in pitch because the lengths of the lever arms are short.

One of the first interactive systems to reach production was the BMC Hydrolastic suspension introduced on the 1962 Morris 1100. Convoluted rubber springs absorbed the road-load inputs from the front wheels which were transferred to the rear via hydraulic connections

On the Citroën 2CV and Morris 1100, the wheels were positioned at the corners for handling stability, which increased the wheelbase: the Citroën's was 94.5 in. (2.4 m) and the 1100's 93.5 in. (2.375 m), each about 65 per cent of the overall length at a time when the wheelbase of most small cars was only 55 per cent. To reduce the pitching moment induced by bump reactions on the relatively-long wheelbase, the front and rear suspensions were interconnected.

In a sense, interconnection between front and rear systems is the simplest possible form of active suspension. If the same spring is shared between the front and rear suspensions on each side (as on the first conception of Citroën's 'people's car' in 1936), any pitching moment caused by deflection of the front suspension is countered immediately by an equal and opposite moment at the rear, cancelling out the disturbance.

The problems start because the centre of gravity is never on the line joining front and rear suspension pivots, so it causes additional moments, which are usually described as dynamic weight transfer. These are not resisted at all by freely-mounted, linked springs, so the interconnection must be compromised to reduce the excessive squat and dive that otherwise would cause the car to hit its bump stops at the rear when moving off and at the front when braking. Citroën achieved this by enclosing the main springs in a floating cylinder mounted between two auxiliary springs which reacted against the body at each end.

The Hydrolastic system approached the problem in the opposite manner, using hollow rubber springs operated by a working fluid that, under certain conditions, was allowed to flow between the front and rear suspension units. This displacement of fluid distorted the shape of the rubber springs and, therefore, changed their stiffness in proportion.

What was really clever about this system was that the two springs on each side worked in series over bumps and in parallel under cornering roll, giving a very novel form of dual rate. Although the small-bore, interconnecting pipes modulated the flow and prevented a simple 2:1 ratio between roll and bump stiffness, the differences still offered a considerable advantage. The design of the internal damping valves, which controlled the fluid flow as well, also responded to the velocity of the input, giving variable damping.

As well as being very susceptible to weight transfer, front/rear interconnection reacts very unfavourably to unequal static loads and causes excessive tail sag when rear seats are filled with passengers or luggage is carried in the boot. For tuning the pitch frequency to avoid unpleasant resonances, the 1100 had twin, steel torsion bars inside its rear subframe. There was also a conventional, rear anti-roll bar to reduce the natural understeer of its front-wheel drive.

The advantage of interconnection endowed the Morris 1100 with an amazing ride and handling balance for a car of its class in 1962. Sadly, the considerable potential of the system was never developed fully, and its cost and complexity finally drove it into history when the Maestro replaced the Allegro in 1983.

The story at Citroën was much more successful. Their adventurous, hydro-pneumatic suspension, introduced in 1953 on the rear of the Big Six Traction Avant, was carried over to the front and rear of the DS19 in 1955. It did not use interconnection, other than to provide self-levelling under load, and was only adaptive in static terms under the influence of pressurized, hydraulic controls which changed the spring rates and damping to provide a near-constant ride frequency.

Probably, it stimulated Alex Moulton into filing his Hydrolastic patent and led directly to many other hydro-pneumatic systems, notably from Mercedes-Benz for the 600 in 1964 and later S-class models, and from a German component supplier called Langen AG of Dusseldorf in 1965.

Non-adaptive but self-levelling air suspension dates back even further to Firestone experiments in the 1930s, which eventually resulted in air springs being fitted to Greyhound buses in 1952. Air suspension was also adopted by Cadillac in 1957 but dropped three years later.

The system used by Citroën (and Langen) employs spheroidal steel containers in which a rubber bag filled with nitrogen under pressure is compressed by the action of the working fluid (mineral oil). Various valves in the system control the flow and pressure of the fluid to provide a constant, static ride height regardless of load, damping of the wheel movements, and a variable rate for the springs caused by changes to the effective area of the convoluted rubber diaphragm separating the fluid from the gas.

However, the Citroën system is actually prevented from responding to rapid changes in wheel position and cannot really be considered as an active system in the same way as Hydrolastic. The Langen system, on the other hand, extended Citroën principles much further towards an active system by interconnecting the hydraulic lines front to rear. Unlike Hydrolastic, the connections were made diagonally and both front and rear spring units were mounted at the rear.

The advantage of this approach was that, although the stiffness of the springs in roll was less than in a non-diagonal connection, the diagonal pitching moments were resisted better. Langen spent several years developing their system for production and fitted prototype units to several cars, including a Morris 1100, a Ford Taunus 12M, a Citroën DS19 and a Mercedes-Benz 220S, but it was never adopted as original equipment.

No-roll systems

The first, truly-active suspension, stabilized to eliminate pitch and roll, was developed by Automotive Products in the early 1970s. A prototype Rover 3500 P6 was used for some very convincing demonstrations at the time, and a similar system was fitted to a Ford Granada research vehicle for assessment in 1974.

The AP system was an extension of the diagonally-interconnected Langen principles, used in conjunction with a primary main system, several motion sensors and a fast-acting, high-capacity hydraulic pump. Unlike the Citroën hydropneumatic system, where time delays are incorporated to prevent any fast reactions influencing ride-height control, the AP suspension was designed to provide the fastest dynamic responses possible.

Control of the system was by a mechanical pendulum valve, designed to match the normal behaviour of the vehicle exactly. Therefore, any transient pump signals were filtered out, but any change in the steady-state conditions caused fluid to be pumped to whichever strut needed to be jacked or relaxed to control the body motion.

The early AP systems generated self-levelling under inertia loads to provide virtually zero-roll and zero-pitch characteristics. These controlled the wheel camber angles so well that cornering speeds were between 10 and 15 per cent faster than in a similar car without the system. However, the system operated by generating a displacement error that was corrected, all of which took time. The double-edged sword which killed the concept before it reached production was that the faster the response time, the more the ride was upset.

LTP 692

The benefits of zero or negative roll characteristics are clearly demonstrated by these two Volvos. The car on the right has conventional suspension, while that on the left is computer-controlled to resist body roll

Lotus effectively removed all these constraints in their active system by replacing the spring-and-damper units with irreversible, double-acting, hydraulic actuators. By measuring instantaneous loads and displacements of each wheel and processing the data to control the actuators, the system became a sophisticated set of microprocessor-activated synthetic springs.

The original stimulus for the Lotus system stemmed from the amazing aerodynamic developments which took place in Formula 1 between 1977 and 1982. Aerodynamic downforce, generated by aerofoils and skirts, increased by a factor of three to a level well in excess of the car's static weight (a car weighing 1650 lb/750 kg, for example, typically generated a purely-aerodynamic downforce of around 2645 lb/1200 kg at 150 mph/240 km/h). As the loads were all applied to the chassis and transmitted to the tyres through the suspension, the demands on the springs and dampers were extreme. Instead of merely coping with 750 kg mass, the suspension had to handle close to 4400 lb (2000 kg).

To compound the problem, the adhesion provided by these forces allowed cornering accelerations of over 3.5 g. It was all well beyond the limitations of conventional springs and dampers and, although skirts were totally banned in 1982, reducing the downforces to peak levels of around 1200 lb (550 kg) or only 70 per cent of the car's mass, Lotus went ahead with their research.

The first test car was an Esprit, followed by a T92 Formula 1 car. They both used the most advanced and sophisticated system of suspension control ever tested. It was based on precision, signal-conditioning units designed by a group of advanced-electronics engineers at the Cranfield Institute of Technology, working with Texas Instruments in nearby Bedford. The system employed digital control of analogue inputs to maintain hydraulic pressure, tailor the dynamic attitude of the car at speed, and monitor the health of the system at all times. The exceedingly-complex systems converted forces, accelerations and displacements to co-ordinates that were relayed to the hydraulic servo valves, after being compared with actual positions and processed in a fast-acting feedback loop.

The result on the Esprit was dramatic. Control of body motion over bumps, and the complete absence of roll or pitch in transient manoeuvres, provided feelings of security and precision that made a quantum leap from even today's high standards at Lotus.

At the time, the active-suspension cars were more valuable to Lotus as a development tool rather than a serious project for future production cars. They offered the facility of altering stiffness, damping, attitude changes, self-levelling characteristics and all the complex interrelationships between these and other parameters from the cockpit while the car was in motion. Eliminating the time delay required to change parts or recalibrate suspension settings between subjective tests transformed the assessment process completely. It gave the development engineers time to reach better performance compromises and allowed new investigations to be made into the various possible strategies for future adaptive systems.

The fruits of all that innovative work, and the subsequent development that has taken place at Lotus since 1982, will soon reach commercial production even with General Motors in control (following their takeover of Lotus early in 1986), although there have been some well-defined statements on autonomy for the company and the honouring of existing contracts. Whatever the future holds for Lotus, there can be no doubt that other manufacturers will take a very long time to catch up with their considerable lead in suspension design.

Other active-suspension systems

The fundamental difference between an active and conventional suspension is that active systems provide independent treatment of road-induced forces and body inertia forces. Therefore, they have the ability to eliminate the traditional compromise between ride and handling characteristics.

In most of the systems being investigated, road disturbances are treated 'passively', as in a conventional system, but with much softer spring and damper settings. Body forces generated by cornering, braking and accelerating are controlled by the 'active' elements in the system, which resist roll and pitch in both the transient and steady-state conditions to keep the vehicle level and hold the road wheels as upright as possible for maximum handling and roadholding performance.

To achieve this functional independence, conventional coil springs are replaced by hydro-pneumatic struts and gas springs incorporating damper valves. Although each front strut and gas spring assembly supports a front corner of the vehicle in the usual way, it is connected to the lower portion of each diagonally-opposite rear strut as well. The upper section of each rear strut is connected to a common, rear gas spring, relative piston areas being used to control the roll-couple distribution, which is independent of ride frequencies.

Ride-height sensors operate self-levelling valves that increase or decrease the hydraulic pressure in the struts to compensate for any load condition, with either electronic or viscous-damped delay mechanisms to allow normal wheel movement on the road. Front-strut pressures are also controlled by fast-response, inertia sensors or valves that can differentiate between bump movements and roll or pitch inputs, responding individually in an optimized manner. The first active-ride systems developed by Ford had mechanical controls, but the potential offered by electronic sensors and on-board computers has resulted in a new generation of equipment.

As a first step towards production with active-suspension systems of this type, the Lincoln-Mercury Division of Ford in Dearborn has developed an electronically-controlled air suspension that was introduced on the 1984 Continental Mark VII models. It is a very advanced, self-levelling installation that provides a constant ride height and consistent handling regardless of load, as would be required for an active system, but without the dynamic controls to restrict roll and pitch. Much of the hardware and control strategy could be applied to an active system.

Because constant suspension travel is always made available through the self-levelling mechanism, spring rates are between 32 and 38 per cent lower than would have been practical for the same car with conventional steel springs. Earlier experimental work on an active control system for a European Granada provided effective operation with spring rates that were 20 per cent lower than standard at the front and 45 per cent lower at the rear, which illustrates the degree of potential ride improvement offered by advanced systems of this type.

The Lincoln system comprises four air springs supplied by an electric air compressor, a single-chip microcomputer control module and three electronic, Hall-effect height sensors mounted at each front corner and over the rear axle. Their linear signals are processed along with dynamic signals that measure the suspension frequency by counting the vertical movements over a time base of 45 sec. Included in the computer logic are certain fundamental assumptions, like not to make any height adjustments when the brakes are applied, and to start

Lotus first tried their advanced active-ride system on an Esprit coupé. The car has good anti-roll characteristics, but still adopts a few degrees of body lean under hard-cornering forces

The active-ride Lotus corners at higher speeds with no measurable roll and with all four wheels square on the road surface, improving grip considerably

Lotus active-ride suspension is totally controlled by fast-acting hydraulic rams driven by the on-board computer

making height corrections immediately after a door or the boot lid has been opened—but never downwards in case the door jams on a kerb. Output from the computer activates either the electric compressor or vent valves, as required.

Apart from meticulous suppression of unwanted electrical inputs, the Lincoln system also has a novel 'watchdog' circuit to protect its functions and the vehicle's battery when parked for long periods. To prevent the computer becoming confused by the ignition being switched on and off several times (as with a difficult cold start), its power supply is arranged to remain active for one hour after the ignition has been switched off, after which time the microprocessor shuts itself down until the ignition is switched on again.

Adaptive suspension damping

As the dynamic performance of suspension systems has improved progressively in recent years, the search for ideal damper characteristics has intensified. Although conventional, velocity-conscious units can deal competently with a wide range of operating requirements, the increased understanding of wheel control has highlighted the need for some form of 'intelligence' in the dampers that can discriminate between different damping requirements, even when they occur at the same wheel velocities.

Much softer springs with variable rates can be specified using electronic height controls to give considerable comfort and handling benefits

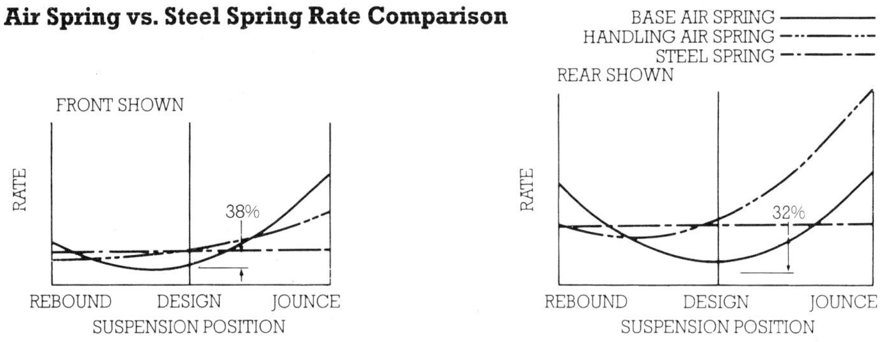

Currently, research engineers in Europe are working on controlling damper logic through an on-board microprocessor fed with inputs from a variety of sensors and driving adjustable valves built into relatively-conventional units. Although alternative damping methods have been explored extensively, so far none has proved to be competitive with the energy density of the hydraulic unit.

The most-important and ever-present operating condition with which the shock absorber has to deal is the road profile, which is varied and random in its character. Frequency inputs to the suspension from this source appear as either irregular or repetitive waveforms, or even as single, sharp disturbances generated by potholes or trench repairs. Many of these conditions present conflicting demands on suspension damping, the primary task of which is to minimize the disturbances felt inside the body while maintaining adhesion between the tyre and the road.

The highest damping forces are required when repetitive road inputs cause a resonance of the sprung mass. Holding down the response amplitudes satisfies both comfort and adhesion requirements. Short, single disturbances, on the other hand,

The effect of an adjustable valve on shock-absorber performance (bottom) is compared here with a conventional fixed-calibration unit (top). Control over the full range of body movement is similar in both cases, but the differences at high amplitudes are significant

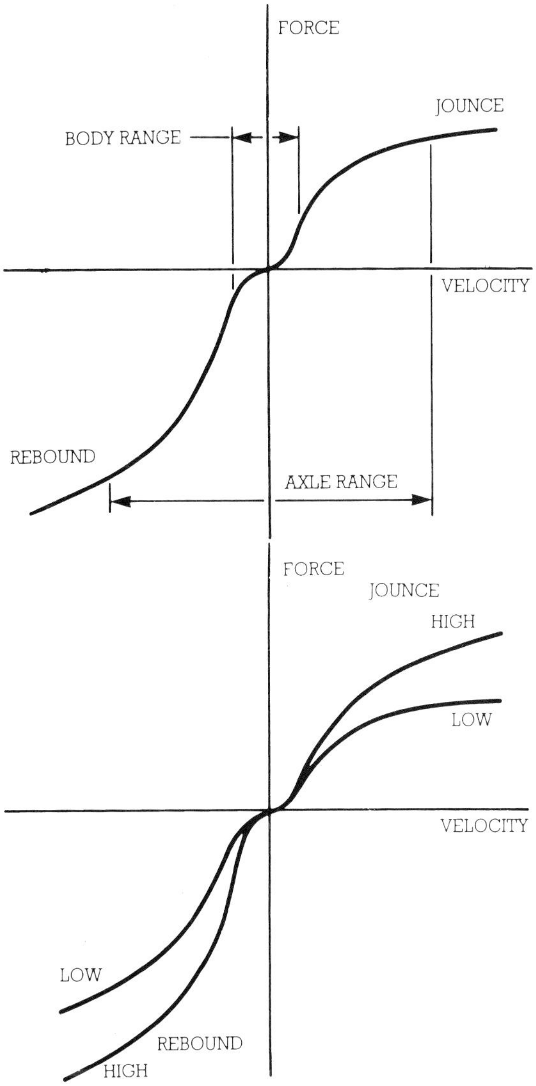

are best controlled with ligher damping forces which have the benefit of transmitting lower forces into the body.

Accelerating or braking causes longitudinal weight transfer that can excite forced oscillations of the unsprung masses (wheel tramp or hop) and cause pitching motions that are particularly disturbing to passengers. Damping forces can only affect pitch motions while the attitude of the vehicle is changing; they are completely ineffective as soon as a steady state is reached and the suspension velocity is zero. They can slow down the rate of pitch change, but then the levels are much higher than those necessary for ride control and consequently transmit larger impulses into the body over road disturbances.

Lateral weight transfer caused by handling or cornering manoeuvres, and the associated body roll, can also be modified by the damper settings. However, as with longitudinal weight transfer, damping can only affect the transient motions, not the steady-state conditions. Damper settings likely to influence handling responses significantly are generally higher than those required for ride control in straight-line running, although small changes can influence stability in high-speed, lane-change manoeuvres.

To approach these conflicts with a 'clean sheet', Ford research engineers started

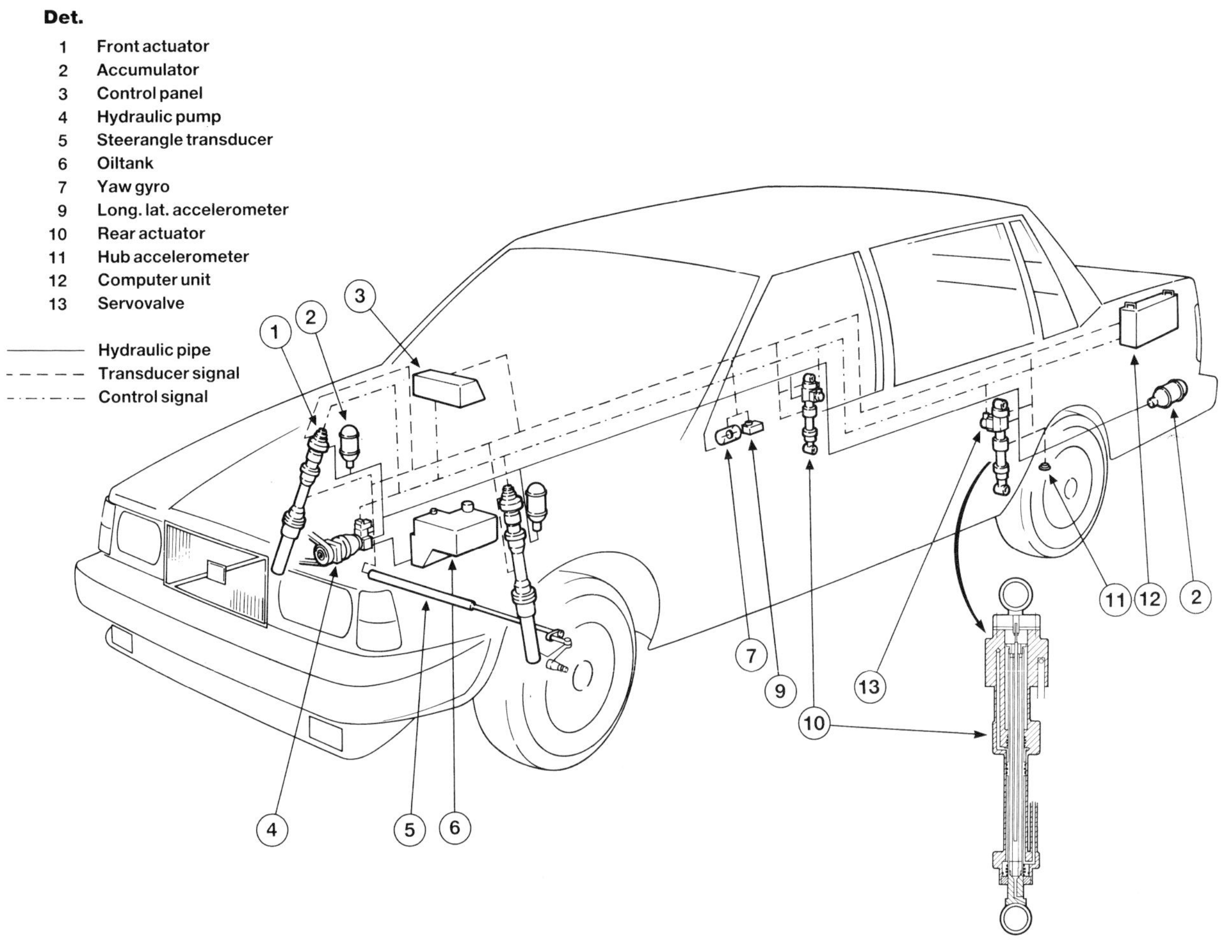

The installation used by Volvo on their experimental computer system required a total of 13 component parts and three kinds of interconnection

with a complete list of ideal, adaptive damping features. Their design used the processing power of an on-board computer to control damping characteristics in jounce (bump) and rebound for different static loading, in response to speed and to changes in the vertical, longitudinal and transverse acceleration of the suspension.

Two such adaptive damping systems have been designed and fitted to a Granada and an Escort for appraisal. In the Granada installation, an inductive displacement transducer was mounted between one rear wishbone and the vehicle body. Signals were filtered and fed in digital form to a microprocessor, which calculated an average wheel-to-body height on the move over a 14 sec sample of amplitudes, comparing it with instantaneous values measured 20 times per sec. Then, control signals were generated to vary the pressure in a vacuum-powered, hydraulic cylinder, which was connected via hydrostatic lines to progressively-adjustable valves in each damper.

Various logic sequences were programmed into the computer for evaluation of the vehicle dynamics. In one typical sequence, damping characteristics were

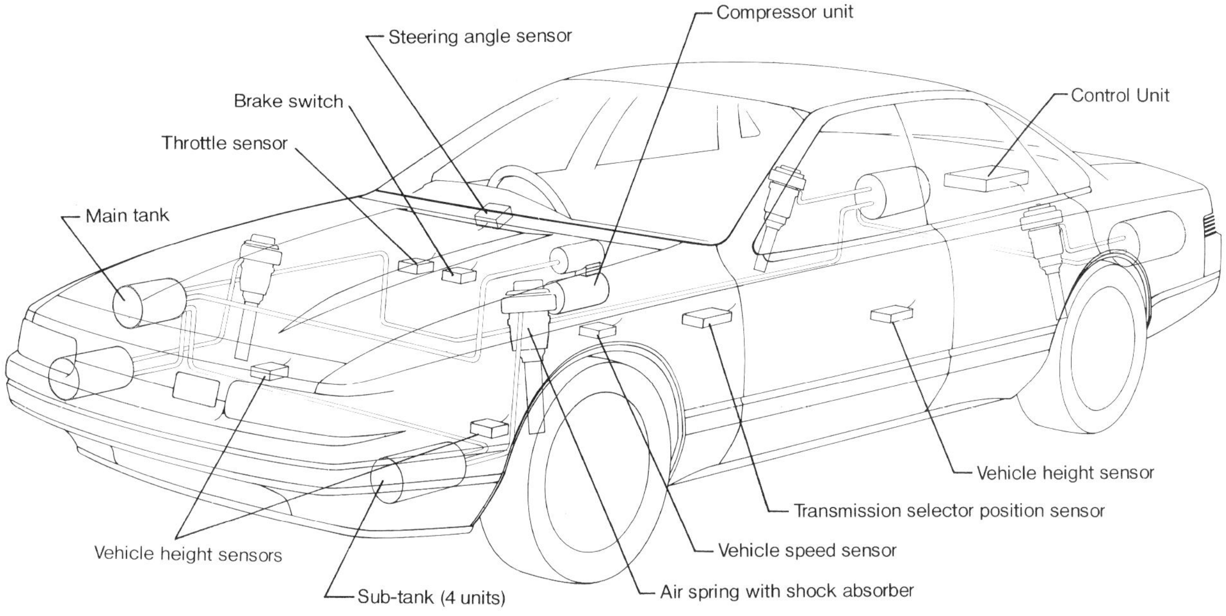

The Nissan CUE-X experimental car was fitted with similar systems to the Volvo, plus sensors for its rear-wheel steering and advanced engine management. Two spring rates, three damper settings and three ride heights were provided

increased for a short, fixed interval (2 sec, for example) whenever the difference between the instantaneous and average ride heights exceeded $\frac{5}{8}$in. (1.5 cm), after which they reverted to the softer settings, provided there were no more inputs beyond this threshold. This design maintained good body control over long-wavelength disturbances, with a generally soft feel to the ride. As the transducer was fitted on one suspension unit only, it also responded to roll, although, in contrast to a transverse accelerometer, some movement was required before any response could be generated.

A much simpler approach was used in the Escort installation, with two self-levelling air struts fitted at the rear only, to supplement the existing steel springs. Inductive height sensors were incorporated in the dust shields of the dampers, which were fitted with two-stage (high/low), electrically-operated adjustment valves for jounce and rebound. The on-board microprocessor controlled both levelling and damping functions through air-pressure signals proportional to vehicle loading and by interpretation of any larger-than-normal, low-frequency body movement.

Soft damper settings were selected for the driver-only or single-passenger condition; hard settings for four people or equivalent loads. Similar logic to that used on the Granada was programmed in so that the system selected harder settings for short periods when excessive body movement was detected. Control of sprung and unsprung masses was well maintained at all times, with ride characteristics that were perceived as softer than the standard vehicle on most road surfaces. In some conditions, it was difficult to detect much difference, however, but the hardware and systems used were only in the first stages of development and operating over quite slow response times.

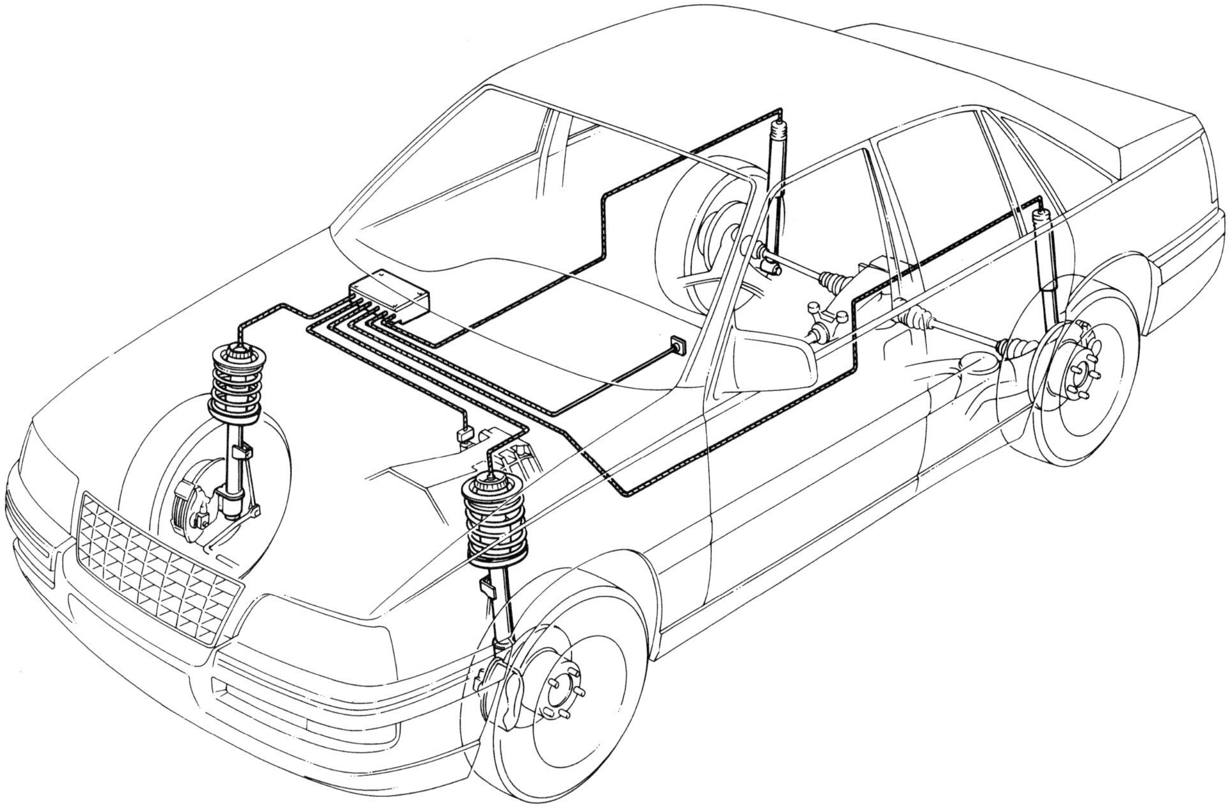

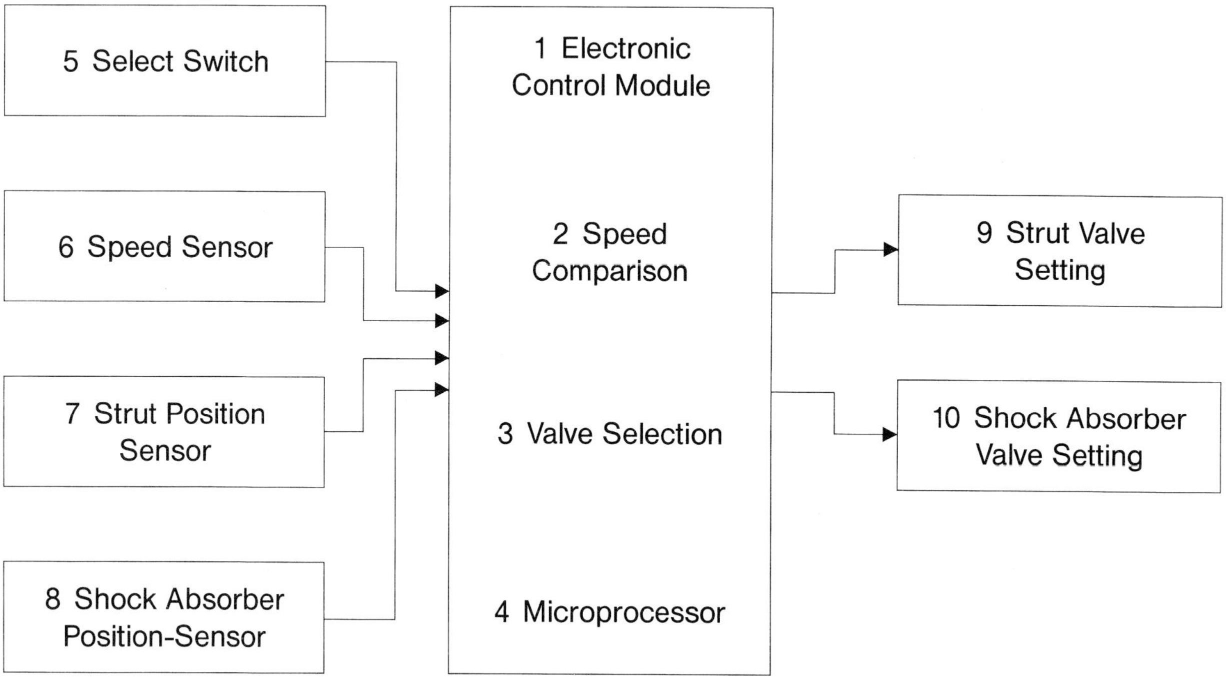
5 Select Switch
6 Speed Sensor
7 Strut Position Sensor
8 Shock Absorber Position-Sensor
1 Electronic Control Module
2 Speed Comparison
3 Valve Selection
4 Microprocessor
9 Strut Valve Setting
10 Shock Absorber Valve Setting

TOP LEFT *Electronic ride control through adjustable damper valves is the latest advance in suspension design to reach production. The system used on the new Opel and Vauxhall Senator is controlled centrally by a microprocessor*

BOTTOM LEFT *There are three damper settings available to the driver, combined with a self-diagnosis programme and speed-sensitive override*

RIGHT *Inside the hollow strut (1) is a high-speed stepper motor (2) which drives the inner part of the valve (3) to vary the flow through ports in the outer (4). Response time is only 300 milliseconds*

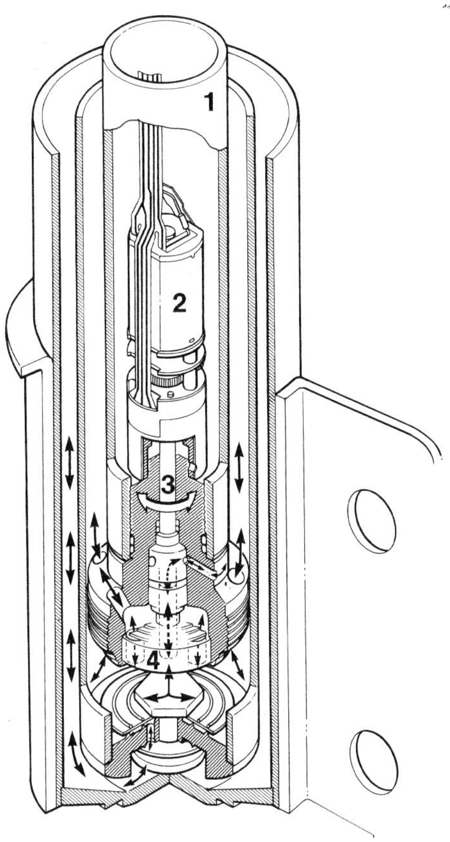

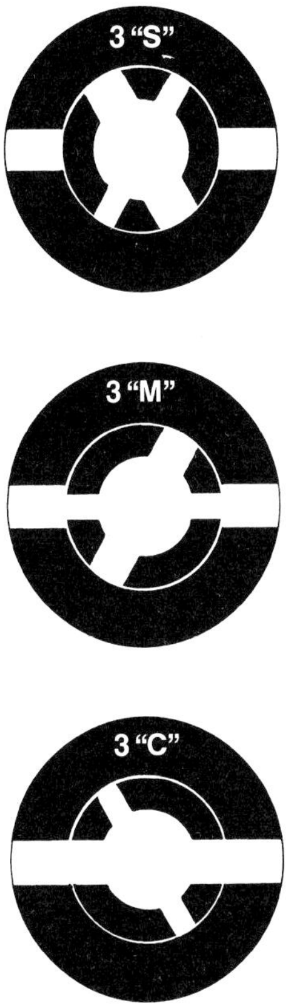

The future potential of adaptive-damping principles, which undoubtedly extend the capability of relatively-conventional dampers, depends on the availability of compact, automotive-quality, high-speed, proportional-control hydraulic valves. Although they do not exist at the moment, major research efforts are being made in this direction which present some exciting prospects for the years ahead.

10 Test and appraisal methods

Performance and durability of suspension components is one of the most critical factors in vehicle design and dynamic control on the road. Throughout the motor industry, advanced test facilities have been developed to ensure that every part of the suspension system meets stringent performance criteria over a long and dependable service life.

Designing mechanical components for a suspension system, like the other parts in a vehicle, begins with a realistic assumption of the stresses likely to be experienced in the most severe conditions of service. Then, the performance of prototype parts is validated and the design developed under controlled, test conditions to reach maximum levels of efficiency, cost effectiveness and manufacturing feasibility.

Although the suspension system appears simple when compared to a power unit or transmission, its function and its effect on vehicle behaviour are complex. Performance criteria for suspension components are based on a combination of engineering experience and input data from scientific simulations and instrumented vehicles operating under actual road conditions. Many of the most severe surface features to be found anywhere in the world are faithfully reproduced on special road systems at the various industry proving grounds, thus saving valuable engineering time.

Multi-channel tape recordings of road input data are transformed into digital computer codes (many hours of testing being condensed on to a single tape) to permit the generation of laboratory test sequences for typical components and complete suspension assemblies. Relationships between test cycles and component life are determined by stress-level analysis, now performed with computer-aided techniques for faster and more reliable production of test-rig control tapes, level-count distributions and frequency analyses of noise and vibration recordings.

Damper performance standards

Although all the springs, links, bushes and insulators of a modern suspension system require meticulous design and development to ensure high standards of vehicle behaviour, the role of the damper has changed completely. It has evolved from a simple set of friction discs to a hydraulic damping piston, and plays an integrated part in providing primary ride characteristics. This has put extra demands upon it, requiring new ways of developing and testing its vital performance standards.

Car manufacturers' proving grounds include test hills, like these used by Audi, as well as special road surfaces built to reproduce the worst the world can offer for suspension appraisal

The often conflicting demands of vehicle ride and handling, and the increased performance of the modern car with its wide-section, low-profile tyres, result in very high peak forces and velocities in the suspension system. Some systems with lightweight, undriven rear axles have recorded vertical wheel velocities approaching 16 ft (5 m)/sec on rough roads, which is several times the peak value traditionally assumed for suspension design. At the same time, on smooth roads, damper-seal friction forces need to be as low as possible to ensure that the ride is not compromised by stickiness in the piston-rod stroke.

Damper performance is a fundamental part of both the handling control of the vehicle and the comfort levels felt by the occupants. The damper settings must meet tight performance and durability targets, even though these components are usually bought in from outside suppliers. Most of the large manufacturers build a very wide range of cars (Ford models, for example, range from the small Fiesta hatchback to the large, luxury Scorpio saloon) incorporating several different designs of suspension system for both front- and rear-wheel drive. Damping equipment specified also covers the spectrum of designs, including gas single-tube, gas-pressurized twin-tube, emulsion-tube and self-levelling types, which are supplied by up to a dozen major European damper manufacturers.

Potholes, sharp ridges, long-pitch waves and other kinds of conditions, including loose gravel, are used in the total test and development process

Lap-top computers like this Epson used by Volvo have become the suspension test engineer's everyday work tool. In this example, the computer is used to programme the on-board active-ride system

Back in the laboratory, recorded traces are analysed and time compressed to accelerate durability testing of suspension components

To ensure that these units meet each company's performance standards, the car makers use prototype vehicles to develop their own individual criteria, which are translated into damper-performance curves, commonly called carding specifications (because traces were recorded originally on small, hard-backed graphs). Durability of the units, in association with the complete vehicle, is determined by running accelerated and general durability tests at various proving grounds. These offer many different road surfaces, including broken concrete, washboard, long-pitch waves, cobblestones, canted slabs, road blocks, chuckholes, railway crossings, square-edged holes and staggered bumps. Test schedules are available with a predetermined severity index that condenses several years' average customer use into a few weeks of testing.

Because of the range and complexity of damper types, it is too expensive and time consuming to cover all aspects of durability on road tests. Therefore, efforts have been made to transfer as much testing as possible to the laboratory. Here, tests can be accelerated more readily, and can be controlled to give more repetition and accuracy, performance being monitored continuously.

Today, damper tests are usually conducted on the latest, computer-controlled machines, some of which can complete more than 3.5 million test cycles in less than four days, equivalent to about 6000 customer miles per day.

One of these new machines, installed at the Ford component laboratories in Essex, took five years to develop and is used to investigate seal and valve durability over a wider spectrum of velocities and temperatures than was possible in the past. It is supported by all the previous performance and durability tests, which are still carried out, including those for strength, corrosion, fatigue and noise level.

Four dampers are mounted together in the machine for simultaneous performance and endurance testing, as they would be on a car. Servo hydraulics under computer control provide vertical motion, while pneumatic actuators apply side loads at the same time, accurately reproducing ride and handling effects from data collected on instrumented cars. Water jackets control the operating temperatures automatically, while computerized recording and shut-down facilities allow unattended operation 24 hours per day.

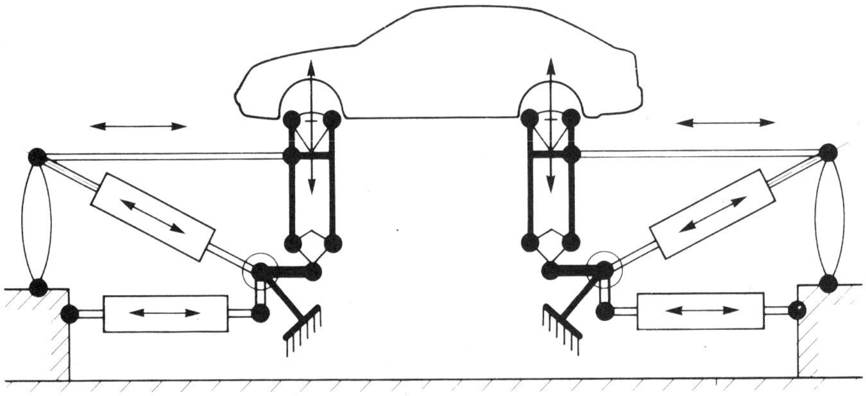

A servo-hydraulic actuator, mounted in the base of the machine frame, applies vertical and axial motion to the dampers. Adjacent to this actuator, and mounted on the sides of the machine frame, are four pneumatic actuators which are coupled together by a cable-and-pulley arrangement and apply a side load to each unit according to a programmable input from the controlling computer. The dampers can be positioned to reproduce any vehicle installation angle between 0 and 30 degrees from the vertical, and there is on-line compensation for any non-linear features in the jounce or rebound input, ensuring accurate reproduction of waveforms.

The control computer, which is shared between two other test machines, has 112 kilobytes of operating system memory and stores all data on two 5-megabyte, hard discs. An automatic plotter can present data in the form of force against displacement, force against velocity for a cycle, or a summary of force against peak velocity for a complete performance test over a range of inputs.

With this highly-developed aid, all of a vehicle's dampers can be tested more effectively and the performance and durability standards controlled more accurately. It is just one example of the investment and methodology that has been applied recently to improve suspension systems and their behaviour in production vehicles.

OPPOSITE PAGE *Real road conditions can be duplicated in the laboratory using computer-controlled test rigs that reproduce the stresses and diplacements measured outside. This kind of hydraulically-driven multi-axis test machine is in common use*

ABOVE *Another kind of road simulator is the 'four-poster' shown here feeding road loads into an active-ride Volvo prototype*

Discomfort levels felt in the car

Levels of comfort (or discomfort) and the amount of fatigue experienced during travel are essentially subjective elements of ride-quality assessment. Disturbances, vibrations, noise levels and harshness can all be measured, but the correlation of these readings to what is felt by the occupants has never been straightforward or always realistic. Consequently, suspension engineers have been taking new initiatives recently to develop objective techniques for quantifying ride levels, with some very promising results.

Ride-quality assessments are usually measured by what is described as a subjective system rating, using a team or panel of test personnel drawn from all areas of company management. The procedure involves a process of detailed and precise comparison between vehicles and suspension settings over a wide variety of surfaces with everything rated on a scale of ten. Below the halfway mark is where the average customer starts to notice bad characteristics, but it takes skilled and developed senses to make a distinction of a half rating on some of the finer subjective qualities, such as secondary ride or steering wheel 'fight'.

To record the disturbances felt by car occupants, specially instrumented dummies are placed in the seats and recordings taken over a range of frequencies in three planes at chest, buttock and foot levels

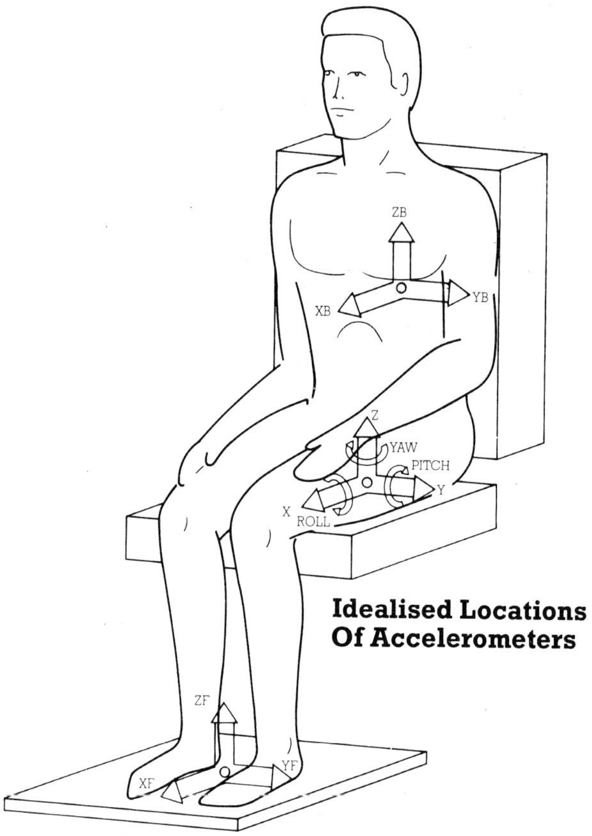

Long lists of average ratings are drawn up from the results to cover all major aspects of the suspension behaviour, from single-bump absorption and pothole shocks to the less obvious, but often more intrusive, qualities like noise, vibration and harshness. Production vehicles, competitive designs and future proposals are appraised fully over all types of road surface, both under controlled conditions in the proving grounds and over public roads. Each development team has its own special location where the characteristics of the surfaces are well known and the resultant suspension behaviour well understood.

Previously, component and vehicle engineers working on ride quality have had little quantitative means of measuring their own achievements. While noise and vibration analyses in anechoic chambers provide valuable development methods for improving refinement levels, the suspension engineer has been restricted in his approach and forced to rely far more on subjective judgements and undefined comments from experts.

Human susceptibility

The medical effects of vibration on the human body have been studied for several years now, with all manner of conclusions having been reached, although most are more concerned with health under work conditions than simple comfort levels from exposure to road disturbances inside a vehicle. Those frequencies which are the most uncomfortable fall either in the relatively-high range, from 20 to 200 Hz (although there is also evidence that fatigue occurs rapidly when a person is subjected to vibrations in the 4 to 8 Hz range), or in the very-low band below 0.75 Hz, where dizziness and motion sickness can result.

Generally speaking, it seems that vertical frequencies in the same range as we experience when walking—between 1 and 2 Hz—are the most comfortable; lateral or fore-and-aft frequencies in the balance mechanism of the inner ear should be of similar values. Therefore, natural frequencies of suspension systems are normally designed to fall within this range, although the total analysis of vibrations will also contain higher-frequency—but lower-amplitude—elements derived from tyre resonances (50–100 Hz) and the compliance bushes (15–50 Hz).

In addition to vibration frequency, measurements of human susceptibility have also demonstrated that the duration of the exposure and its strength (as measured by the root mean square of the acceleration amplitude*) are also critical factors in recorded fatigue levels. Recommendations by the International Standards Organization specify vibration levels likely to cause 'fatigue-decreased proficiency' as a series of contours based on these terms.

Earlier work on ride quality has demonstrated that the root mean square of the vertical acceleration at the vehicle's seat cushion correlates well with subjective ratings determined by appraisal teams. Therefore, the first generation of ride meters measured this quantity over a specified time, and had the facility for using different frequency filters if desired.

Recent developments have extended the scope of this basic approach to the much more complex system shown in the diagram opposite, where human susceptibility to accelerations is measured about each of 12 separate axes. Three linear accelerations (in the X, Y and Z planes) are recorded at chest, buttock and foot levels, together with yaw, pitch and roll accelerations at the seat cushion, for subsequent analysis against amplitude and frequency weightings.

From a weighted combination of these 12 RMS measurements, an overall ride value can be computed which, so far, has correlated very closely with the subjective ratings on ride quality. It provides an extremely useful tool for making rapid comparisons between vehicles or suspension settings. For an immediate read-out of

*Root mean square (RMS) values are derived by taking instantaneous measurements of acceleration over a complete frequency cycle, squaring them and taking the square root of their average.

overall ride values, a simplified, six-channel meter has also been developed to supplement the full analysis of 12-channel recordings on a laboratory computer system.

Self-levelling

Ride comfort is often handicapped by the need to maintain adequate suspension travel in the fully-laden, static condition, which can demand stiffer spring rates than the optimum for driver-only or two-occupant operation. One solution is to use some form of progressive springing that increases its stiffness with load, and another is to add a self-levelling system that restores ride height automatically whenever static wheel travel is detected.

Progressive-rate springs may be wound from tapered wire or provided by second-stage assistance from the latest type of bump stops or snubbers, which are made from new-generation plastics and designed to operate very early in the wheel-travel sequence. Various types of self-levelling suspension (see Chapter 9) are able to maintain ground clearance and keep wheel geometry closer to its optimum for more consistent handling.

11 Rear-wheel steering systems

Although it has only recently entered general production on several 1987 Japanese models, there is nothing particularly new about rear-wheel steering. It is used on dumpers and forklifts for manoeuvrability, on tailwheel aircraft and supermarket trolleys. However, the idea of rear-wheel-steered cars has haunted designers for decades. In 1946, one of the most original concepts was built into a prototype, appropriately called The Crab, by Freddie Dixon and Tony Rolt (see Chapter 5). Actually it was a four-wheel steering system that effectively pulled the inside wheels together in a turn at the same time as it pushed the outside wheels apart.

In the late 1960s and early 1970s, attention turned to rear-wheel steering for easier parking. Audi engineers made one of the first 100 saloons move almost sideways into a parking bay by adding a steering rack to the rear wheels, the interconnecting link being deactivated for normal driving. The same idea was revived by Mazda in 1982 for an advanced research car that appeared at the Birmingham Motor Show. However, all these experiments were little more than publicity-seeking gimmicks.

The easiest way to build early four-wheel-steering prototypes was to cut and weld two front ends together, as on this Honda Accord special. It led to the production version of the four-wheel-steer Honda Prelude launched in 1987

Genuine rear-wheel steering is a much more subtle development than that. Initially, it was designed as an active part of the suspension geometry for the now-famous Porsche Weissach axle, used on the 928 since its launch in 1976, and it appeared again under hydraulic control on the Nissan Bluebird saloon and MID-4 coupé in 1985. In 1986, it appeared on a new Honda and has been investigated extensively by all other major producers.

What is not often appreciated, however, is that rear-wheel steering has actually featured in the predictable handling of many cart-sprung, live-axle systems used before irs became so fashionable. Leaf springs wind up under power, which has little effect in a straight line, but it causes asymmetric effects that generate rear-axle steer when traction is lost at the inside wheel by body roll in a turn.

To understand what rear-wheel steering is able to achieve, it is necessary to go back to the basic principles of vehicle dynamics and modern suspension design. The first point to grasp is known as the Circle of Forces—a fairly-accurate theory which states that a tyre footprint, for any given inflation pressure and vertical loading, usually reaches the same limiting friction force regardless of the direction involved. Basically, it means that tractive or braking forces decrease the lateral cornering

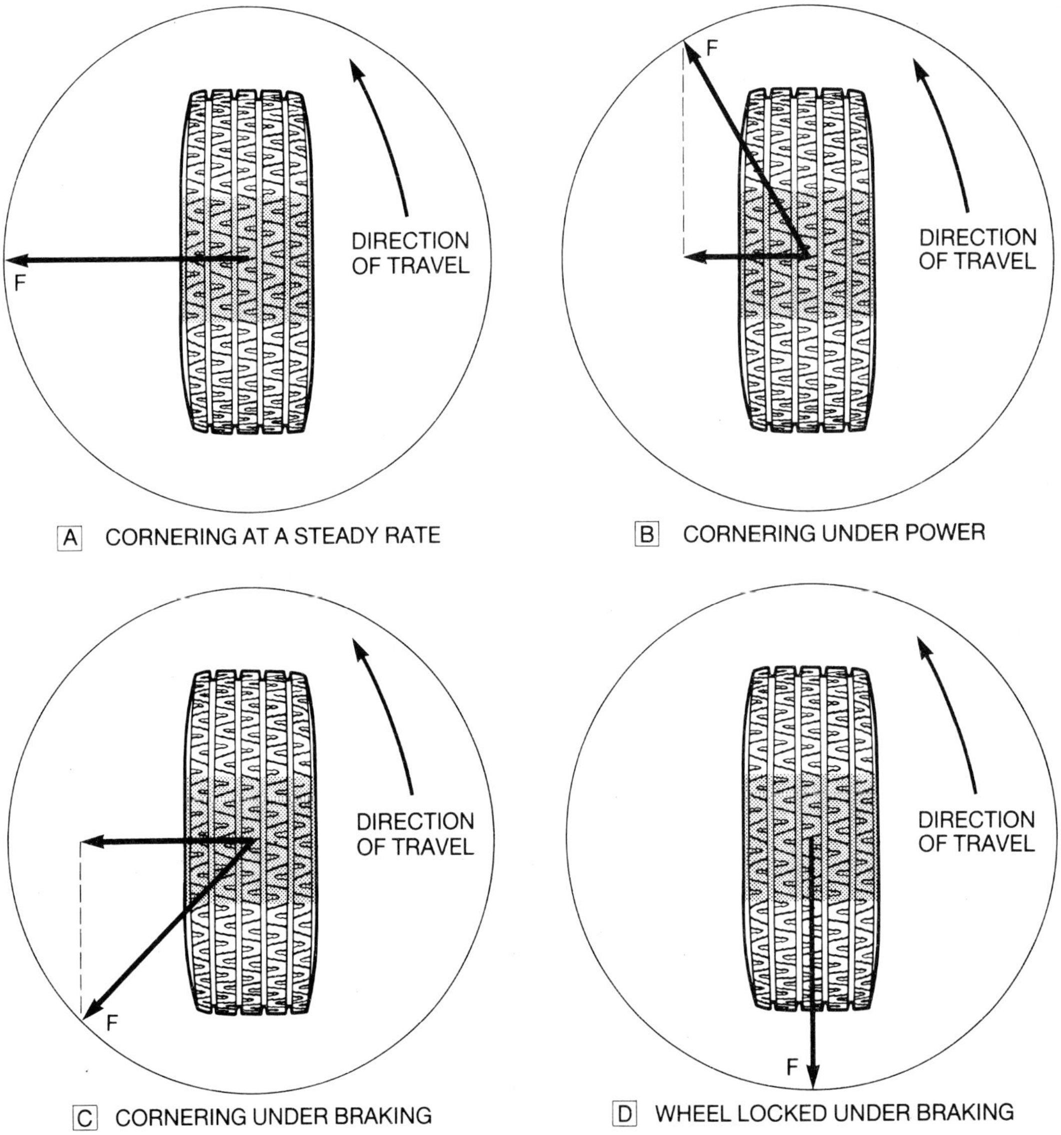

The Circle of Forces is fundamental to tyre behaviour. The frictional force available has a limit (F) that takes no account of direction, so less cornering power is available under power or braking than in a steady state

grip, but that the resultant force vector always has the same value. After making due allowance for the dynamic behaviour of the tread, a tyre will skid forwards at the same limiting force as it skids sideways, and at the same resultant force in any direction between the two extremes. It is the main reason why locked wheels cannot be steered.

Conventional semi-trailing rear links have enough bush compliance to turn out under braking forces (P), causing the more loaded wheel on the outside of the bend to induce oversteer

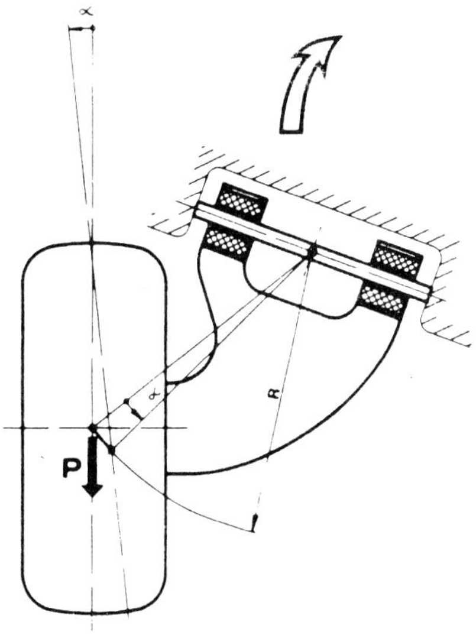

That said, the next principle affecting handling concerns slip angles, or the angular difference between the tyre's direction of travel (the path of the wheel) and the direction in which it is pointing. Slip angles in a turn are generated by distortion of the tyre footprint (contact patch) due to the cornering forces which collectively must oppose the centrifugal forces involved for the car to be in equilibrium. Slip angles are designed to be unequal front and rear to give the car its particular handling balance.

In an understeering car, front slip angles are greater than those at the rear. The vehicle is inherently more stable than an oversteering car because the driver has greater control of the lateral forces (it does not require opposite lock to change direction). Typically, at moderate cornering speeds, front slip angles might be 7 degrees, while those at the rear are only 6 degrees. The limit of adhesion for most tyres occurs at a slip angle between 10 and 14 degrees, after which the cornering force available decreases with slip angle as the tyre starts to slide.

Although simple theory states that friction should be independent of footprint area, that is not true for a rolling tyre tread. Factors which reduce the footprint area, such as inflation pressure, tyre loading and wheel camber, also influence the limit of adhesion. Under dynamic cornering conditions, there is lateral weight transfer about the roll axis, and under power or braking there is longitudinal weight transfer about the pitch axis. The amount of this weight transfer alters the tyre loading and, in most cases, the suspension geometry, complicating the issue considerably.

Before tyre profiles began to reduce from 82–85 per cent aspect ratio (section height/width), camber-angle changes were not critical (about 5 degrees positive camber caused only a ten per cent loss of cornering power). Consequently, wishbone systems or parallel trailing arms, which kept the wheels at the same angle as the

RIGHT *On the design used first for the Porsche 928, the braking forces have more effect on the forward pivot so that the rear wheel turns in, greatly improving stability when slowing in a bend*

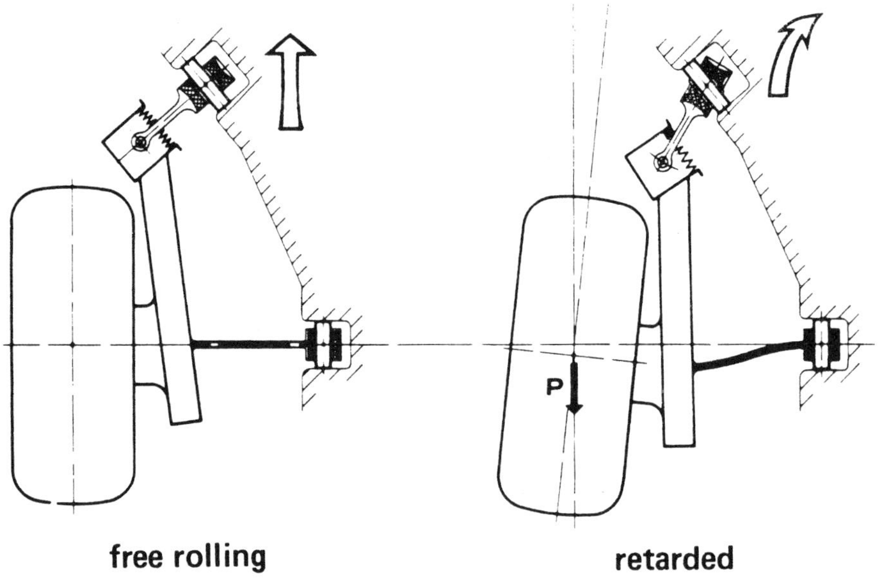

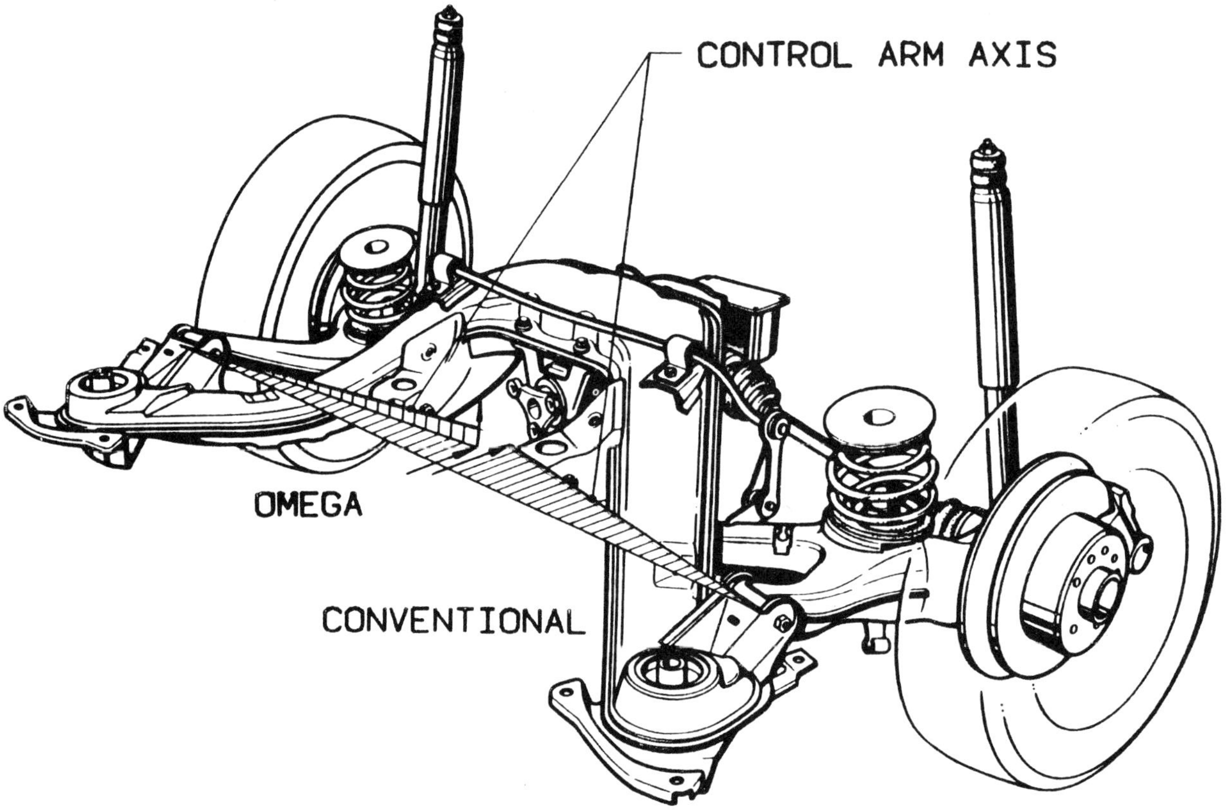

body during roll, were satisfactory, although not as good as rigid beam axles, where the wheels are nearly always vertical to the road surface (the exception being due to compression of the outside tyre wall under lateral weight transfer which led, eventually, to inside wheel lift).

More importantly, however, 80-series tyres provided enough ride comfort for the suspension bushes to be fairly stiff in compliance; the tyre sidewalls absorbed the worst of the high-frequency vibration and harshness from the road surface. In addition, there was enough flexibility in an 80-series tyre to keep most of the footprint in contact with the road during camber changes with suspension movement.

Swing axles, where the wheel is always at right angles to the drive shaft, have the advantage of providing negative wheel camber with spring compression, and the disadvantage of moving into positive camber on rebound. In a corner, when weight is transferred forwards from a rear-mounted swing axle by the driver lifting off or braking, the heavily-loaded outside wheel can swing from negative to positive camber, losing cornering power in the process—as many a driver of a Porsche, Skoda, Dauphine, Beetle or Merc 300SL has found to his cost.

The natural compromise between a swing axle and parallel or trailing links, which have been adopted almost universally for irs, is the semi-trailing arm. It can be arranged to induce slight negative camber at the outside wheel when cornering without the tricky lift-off oversteer of a swing axle. It works extremely well with 80- and 70-series tyres. However, when profiles are reduced to 60- or 50-series, the pivots must be made softer for acceptable insulation from road shocks by adding compliance to the rubber bushes. Overrun or braking forces generate a horizontal couple that turns each wheel outwards by flexing the inboard and outboard bushes, inducing a slight amount of toe-out, which used to be insignificant but now has reached levels of half a degree or $\frac{1}{16}$in. (1.5 mm) at each wheel rim. Although that

LEFT *The compound angles of the new Opel Omega (Vauxhall Carlton in the UK) rear suspension provide increased negative camber during bump and rebound movement and an automatic toe-steer effect when lifting off*

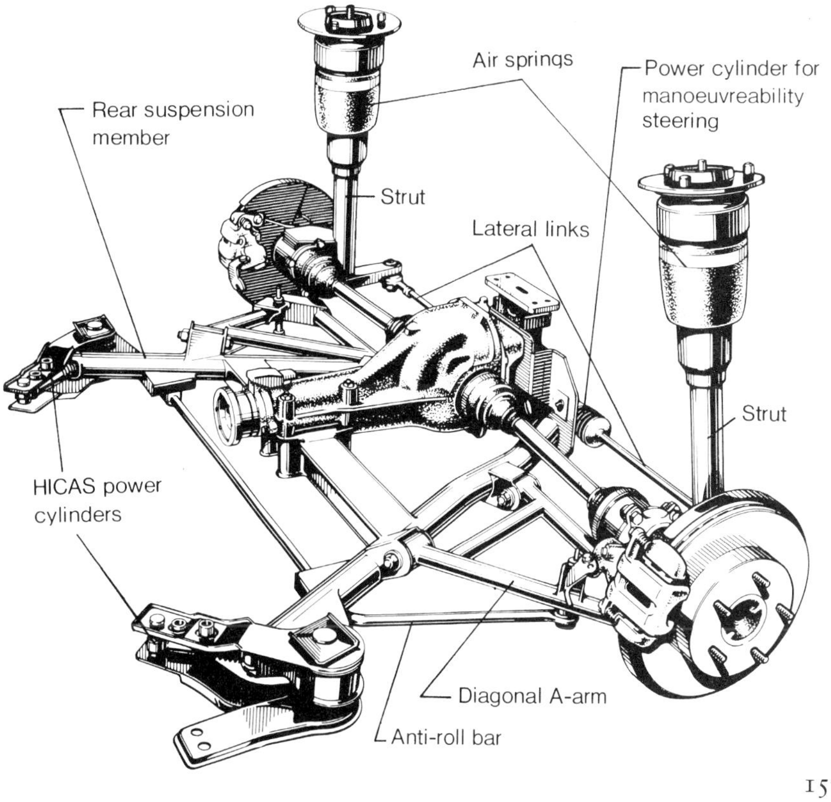

RIGHT *Elements of the Nissan HICAS rear-steer system include a power cylinder for large-angle parking manoeuvres and small hydraulic rams to control the front subframe pivots in high-speed driving*

presents no problems when running in a straight line (it is a symmetrical effect), it increases the slip angle in a turn where the outside tyre is loaded more heavily, cutting cornering power, much like the behaviour of a swing axle, but not so severely.

With pure transverse wishbones (90-degree trail angle), the bushes can be made very stiff in the lateral plane and soft only in the longitudinal direction of bump-induced harshness. However, the flexibility of the wheel under deceleration or braking increases as the trail angle reduces.

When trail angles were around 25 degrees, all these characteristics could be controlled to give safe, predictable handling under all manner of transient conditions. However, trail angle also sets the position of the effective centre of the suspension arm and thus the height of the rear roll centre, so in the search for improved handling there has been a trend towards reducing it. This began in 1977 with the Opel Senator, which nearly halved conventional trail angles to only 14 degrees, followed by Ford, who went to 22 degrees with the Granada S-Pack suspension in 1980 and now run at about 18 degrees on the Sierra and Granada/Scorpio.

To prevent toe changes at the rear, Porsche invented what they call the Weissach axle (after the location of their R & D department), which was introduced on the 928 in 1976. It uses a stiffly-articulated trailing link to prevent deceleration loads reaching the front bush. Instead of the drag forces pulling the semi-trailing arm round to toe-out, they effectively lengthen a spring-loaded joint at one end of the trailing link, keeping the toe changes to a minimum.

Nissan's new system goes one better. Called HICAS (High Capacity Active Control Suspension), it is operated hydraulically by two sets of rams at speeds

The experimental four-wheel steering system developed by Ital Design experimentally for the mid-engined 4 × 4 Incas coupé provides ±34 degrees of normal steering lock combined with ±5 degrees of rear-steer angle

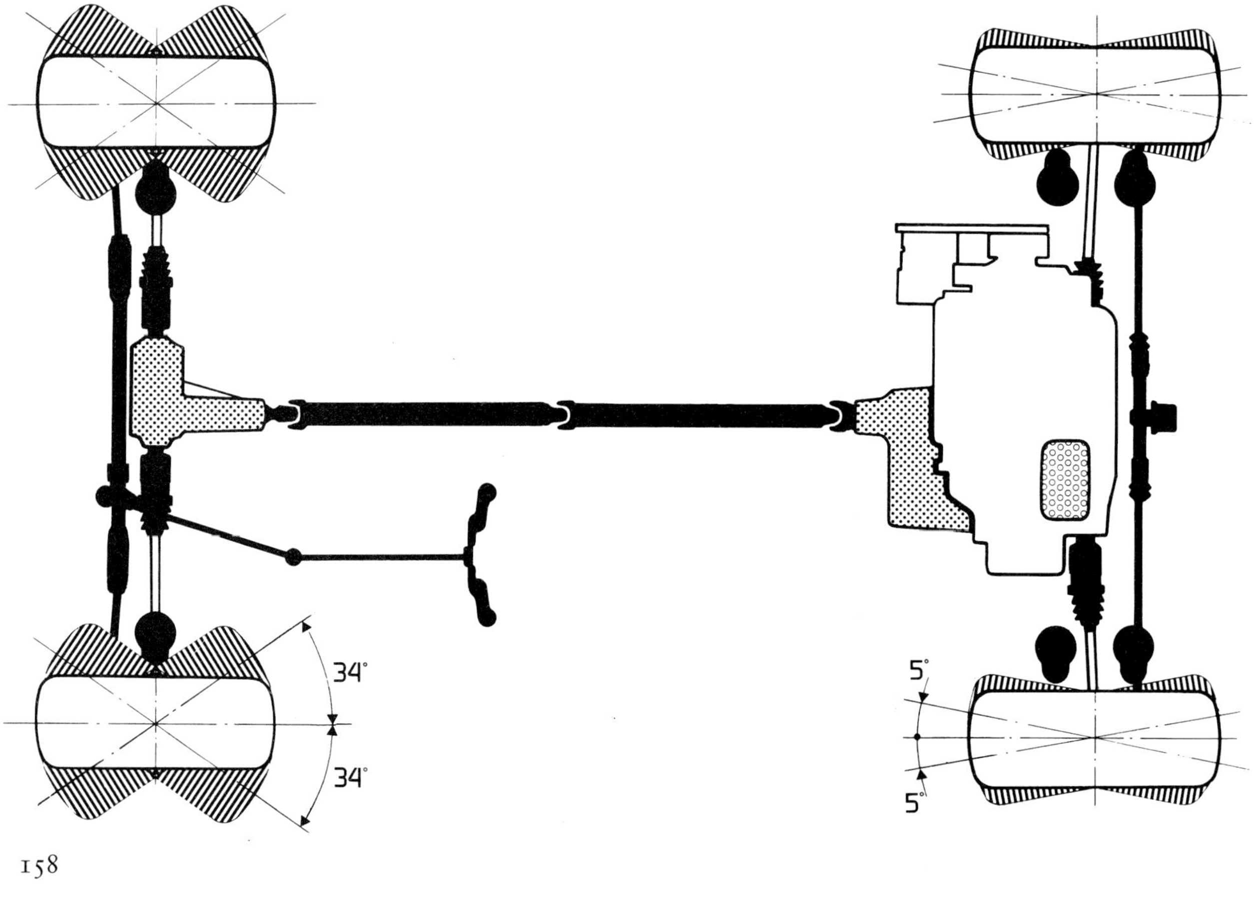

above 20 mph (32 km/h). On the Bluebird, the rams simply 'drive' the rear subframe round on its rubber bushes, but on the MID-4 coupé there are proper steering arms and knuckles, rather like front hubs. When the conventional, front, power-steering system senses high cornering forces, HICAS is activated under microprocessor control to add or subtract up to 0.5 degrees on each side. That means there is a full degree of slip-angle adjustment available at the rear through the rear-wheel steering.

Judged on its own, 1 degree sounds very little (it is equivalent to about $\frac{1}{8}$in./3 mm toe change), but it is enough to counteract all the bush compliance. Cornering power (defined as the cornering force divided by the slip angle generating it) varies by about 15 per cent per slip-angle degree change at normal cornering speeds, which is more than enough to eliminate completely any trace of lift-off instability.

With a typical rear tyre carrying about 660 lb (300 kg) of static and dynamic load under power, a 1-degree decrease in slip angle reduces the cornering force from about 397 lb (180 kg) to 353 lb (160 kg) which, in some conditions, could well reduce it from the threshold of breaking away. Conversely, an increase of 1 degree, caused by lift-off toe-out, could initiate rear-end breakaway.

If the load on the same tyre is reduced by 66 lb (30 kg) to 595 lb (270 kg) by dynamic weight transfer forwards under deceleration, the rear cornering power

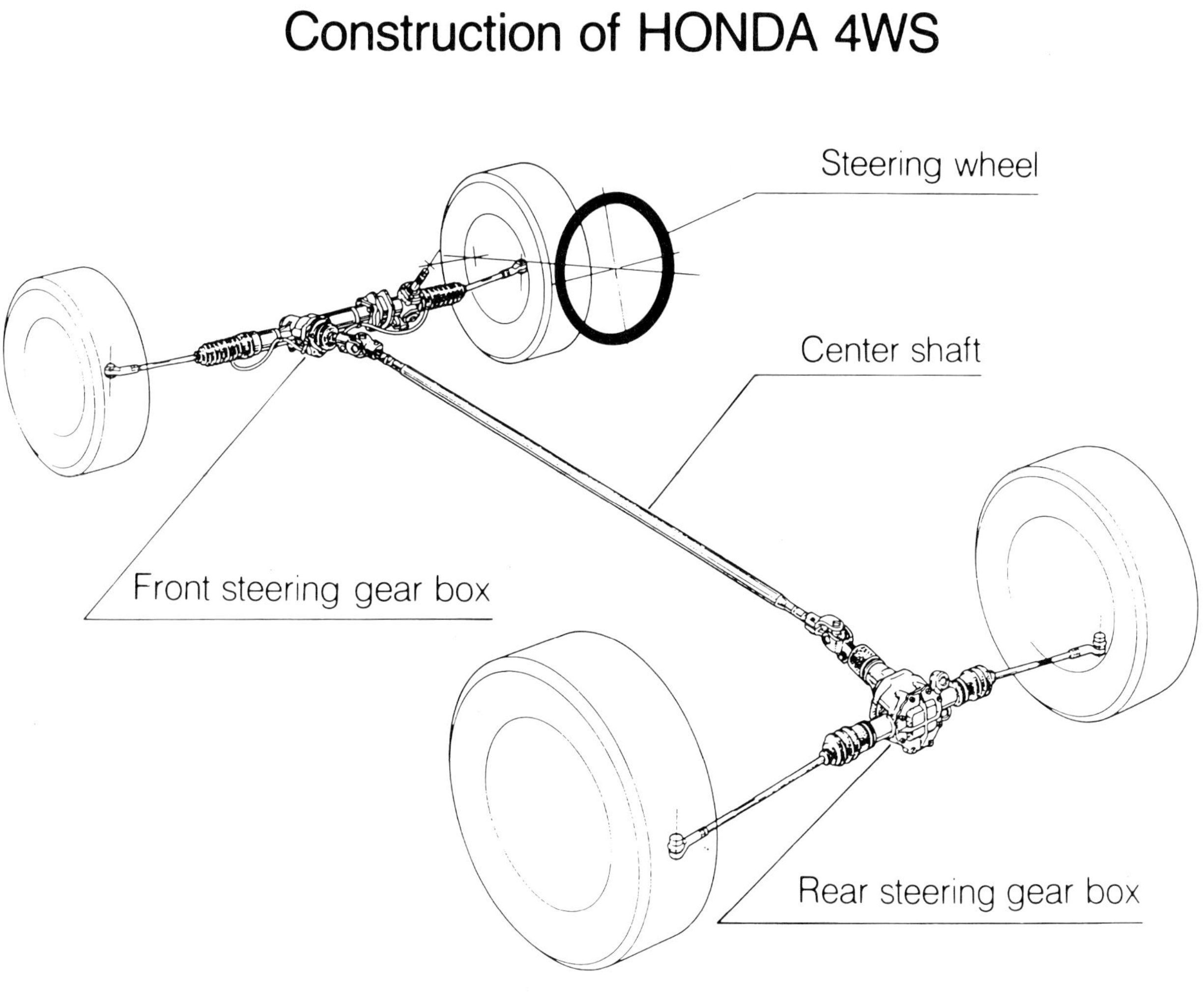

Honda's four-wheel steering system uses a novel shaft-drive rear steering box that automatically changes the direction and amplitude of the rear steer according to the amount of lock applied

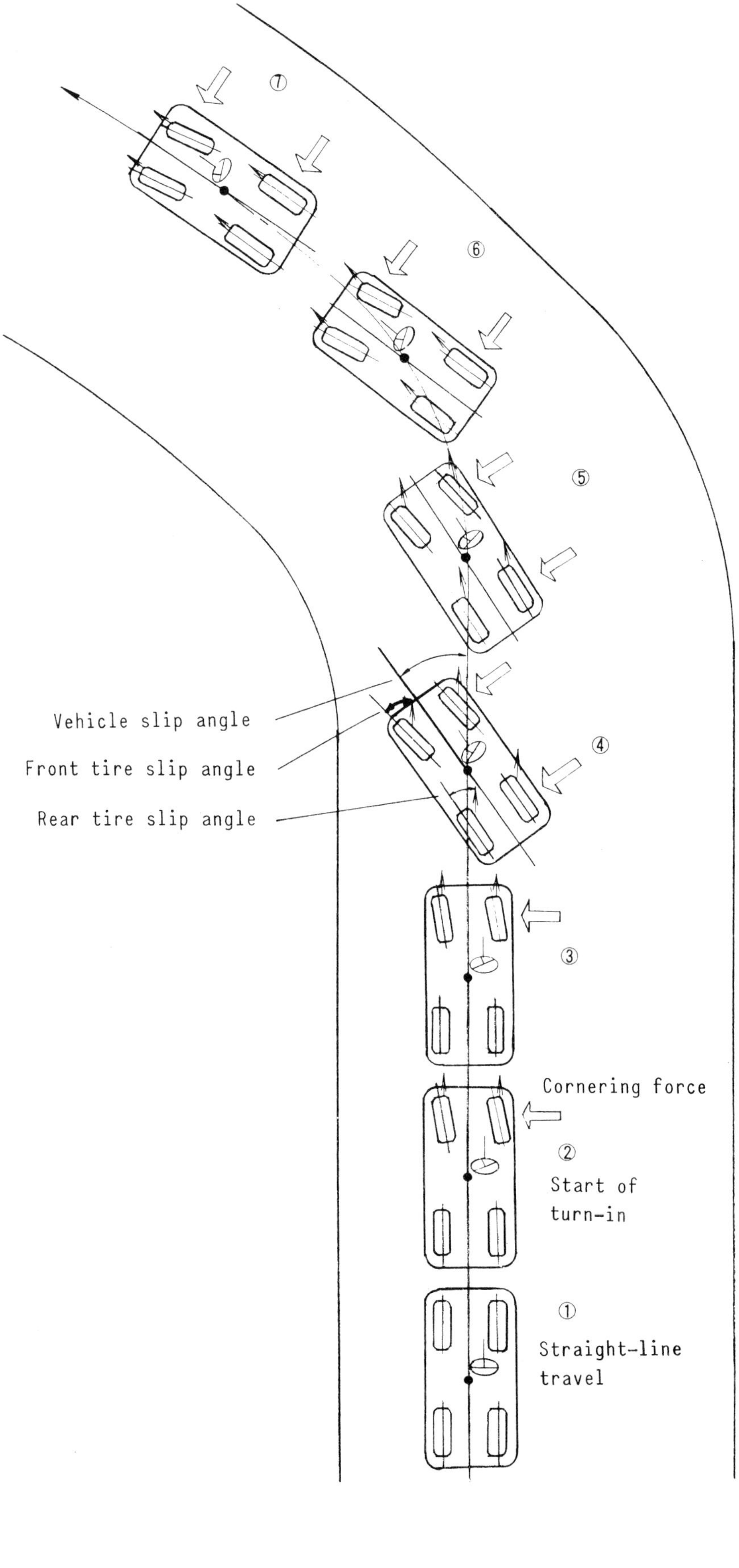
⑦
⑥
⑤
Vehicle slip angle
Front tire slip angle
Rear tire slip angle
④
③
Cornering force
②
Start of turn-in
①
Straight-line travel

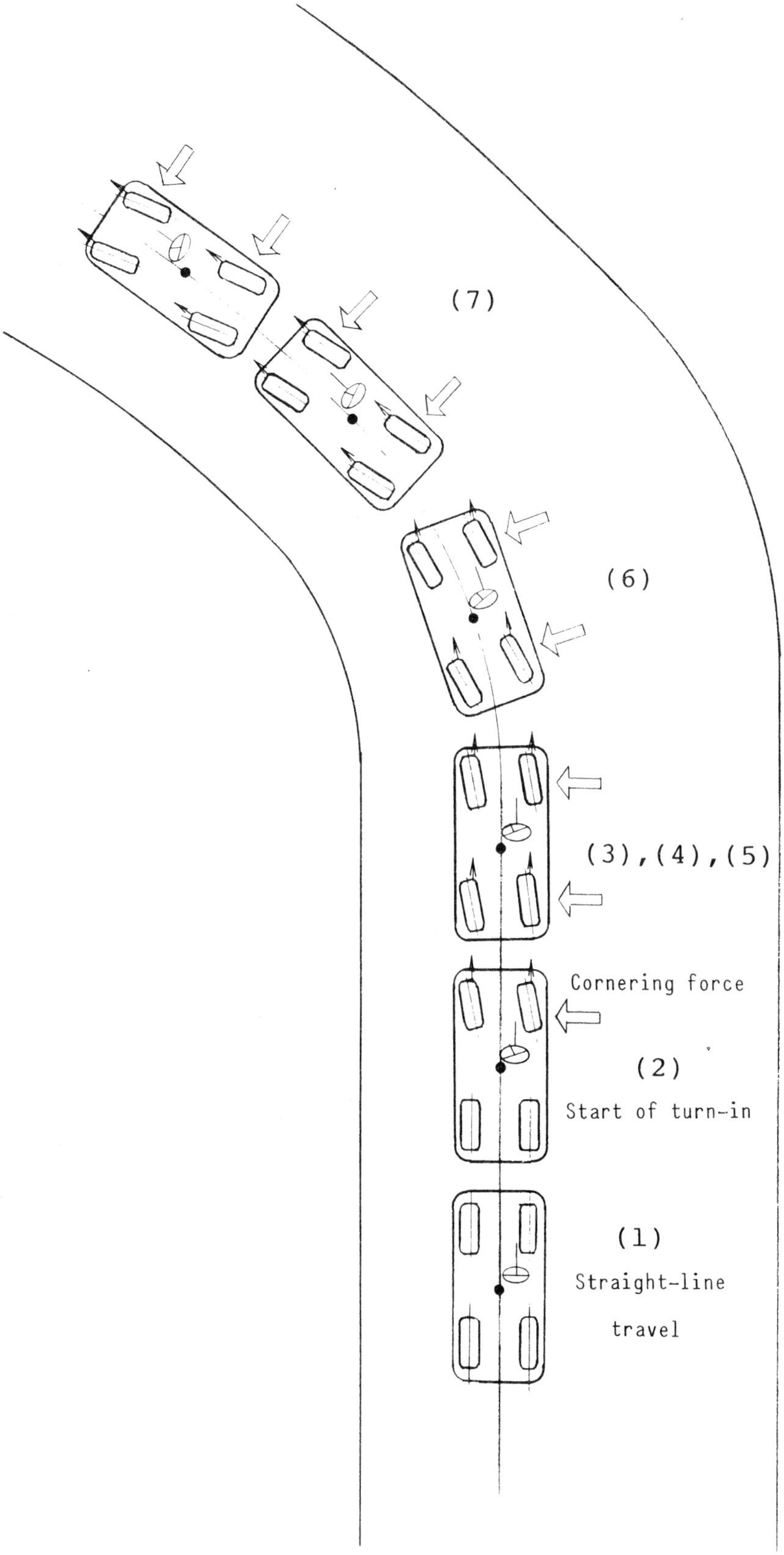

These two sequences compare the behaviour of a typical car in a cornering manoeuvre with normal two-wheel steering (left) and rear-wheel steering (right). The stages 3, 4 and 5 are compressed to sharpen the handling and improve stability considerably

RIGHT *The path traced by the rear wheels in a U-turn manoeuvre can be sharpened with opposite-direction rear-wheel steering to reduce the turning circle considerably*

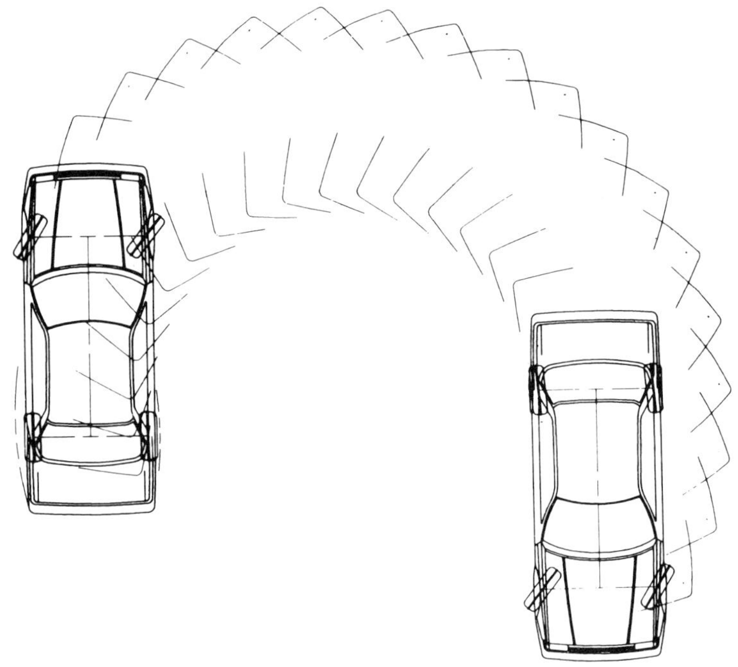

BELOW *The same applies to reversing manoeuvres like this, where the intermediate position is reached sooner and the parked position at the kerb requires less overall length*

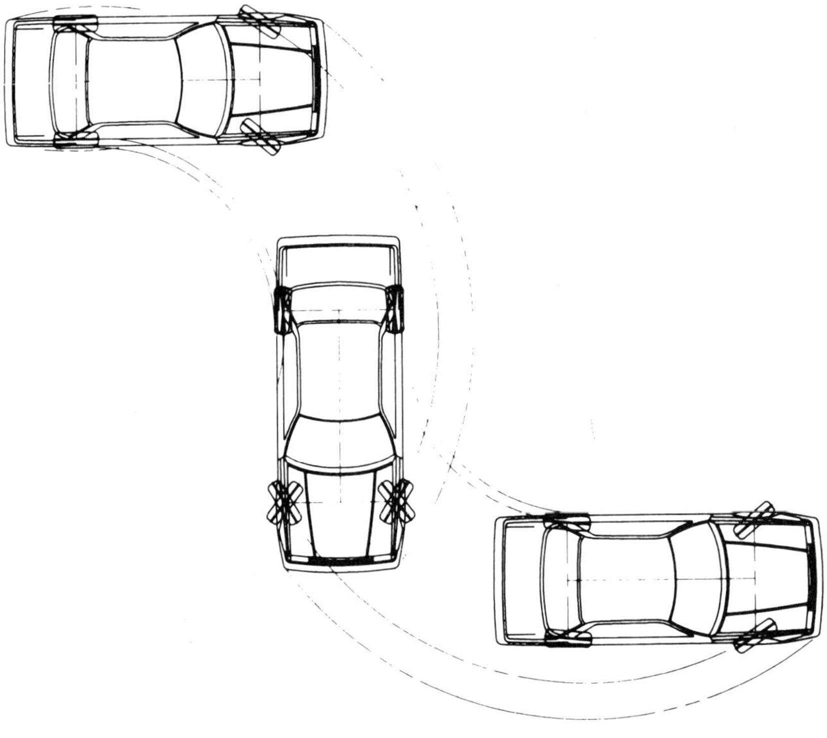

drops and the rear slip angle will rise to over 7 degrees. If the front tyres were running at 7 degrees before the throttle was cut (7 degrees front and 6 degrees rear gives stable understeer, remember), the weight transfer would increase their cornering power by a similar amount, and reduce their slip angle to around 6 degrees. Consequently, the balance is reversed completely, from 7 front/6 rear to 6 front/7 rear, which throws the car from understeer to oversteer, causing noticeable instability.

With the HICAS system, the behaviour is completely different. At the critically-unstable moment, the power-steering hydraulics sense the extra front-end cornering power from the reduced steering effort, and the microcomputer opens the valves to the rear rams so that the rear slip angle is reduced by 1 degree. This cancels out the rear toe change and raises the rear cornering power in step with the increase at the front, keeping the car in balance.

Like all good ideas, HICAS is simple in its effect, but it is complex, costly and critical in its calibration. Nevertheless, as tyre sections come down even further, from 60 to 50 and then 45 per cent, rear-wheel steering will be essential for stability with the suspension geometry wide tyres require and the insulation against harshness needed in their suspension bushes.

12 Braking systems and anti-lock

As an engineering concept, the idea of brakes which are prevented from locking the wheels in an emergency is not new. All airliners and most military aircraft have them; Jensen adapted an aircraft system for the high-performance, Interceptor FF four-wheel-drive coupé in 1966, and Mercedes-Benz introduced a pioneering, Bosch electronic system as an expensive option on their prestige models in 1970.

For the past few years, abs (which, to be accurate, stands for the German term anti-blocker system) has also been readily available on BMWs, Audis and Porsches. It was a standard feature on the 1985 Ford Granada/Scorpio range, and similar systems are optional on the Honda Prelude, Vauxhall Carlton (Opel Omega in Europe) and Senator. From the beginning of 1986, the Ford Escort made low-cost-vehicle history by being offered with a simpler form of abs.

A tyre exerts its maximum grip when it is just on the point of locking. As soon as it stops turning, the coefficient of friction takes a sudden dive, the braking distance goes up and steering control is lost because a static wheel generates no side force. It takes a highly-skilled driver to sense the point of wheel-lock and release pressure on the brake pedal, particularly in the panic of an emergency when reflexes tend to freeze. On most slippery surfaces, the difference in stopping distance between braking with the wheels locked and with the wheels on the point of locking can be as much as 40 per cent. The exception is on fresh, lightly-packed snow or loose gravel, when a wedge builds up ahead of a locked tyre to stop the car more quickly than with abs.

Anti-lock brakes use sensors to detect wheel-lock before it occurs and release brake pressure automatically, allowing the wheels to just keep turning, holding the braking forces at their most effective level and maintaining steering control. In practice, this creates a cycling process similar to cadence braking, a driver-activated technique of pumping the pedal up and down with a rhythm synchronized to excite the natural pitch frequency of the suspension and increase the dynamic load on the front tyres. However, with the benefit of fast-acting, electronic sensors and controls, today's automatic systems operate at much higher frequencies (up to 15 Hz).

Mercedes-Benz were the pioneers who have had the most effect on abs. The Jensen system, using Dunlop Maxaret controls developed directly from aircraft applications, died along with production of the FF coupé in 1972. At the end of 1970, however, Mercedes became the first company to announce an electronically-controlled, anti-skid, braking system for volume-produced cars. Although it was not generally available until the 1979 S-class arrived, it had been under development for nearly 20 years. Currently, well over a million Bosch abs systems have been built.

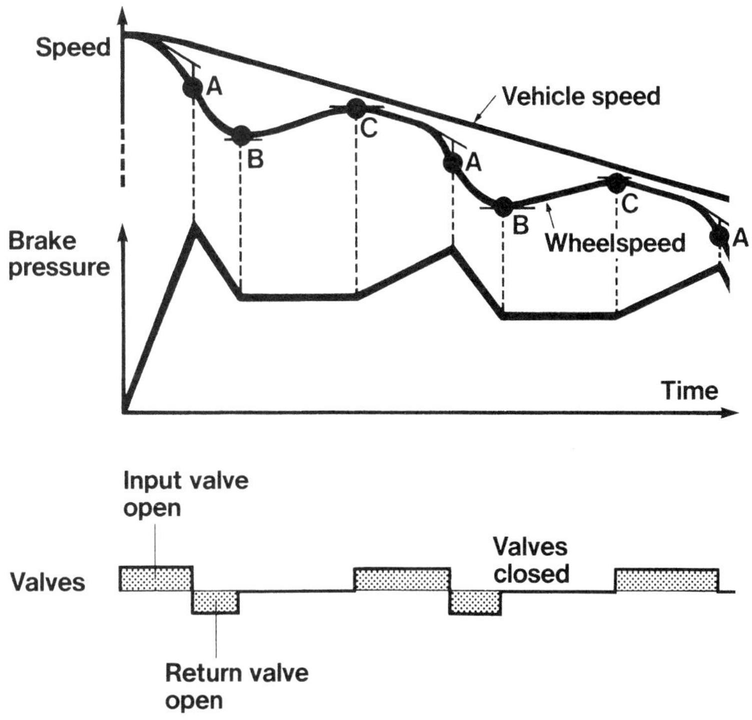

The principle of anti-lock is based on the theory that any braked wheel must be prevented from sliding by modulating the brake pressure in a series of on/off cycles as close as possible to the limits imposed by the friction of the road surface

Because braking is such a life-and-limb affair, each system incorporates a self-test function that simulates a braking sequence and evaluates the signals every time the car is started, checking the most important functions all the time it is moving. Until 1986, such a system was a complex and very expensive piece of equipment, requiring the integration of hydraulic and electronic functions to aircraft standards. It was an area of technology totally dominated by Bosch, who supplied it as standard equipment for Mercedes V8 models, and competitive luxury cars in the BMW, Audi and Porsche ranges.

At the beginning of 1986, however, Ford changed the marketing concept of abs by providing a competitive system from Teves as standard equipment on every new Granada/Scorpio, including the bottom-of-the-range 1800 GL version. To compete, Opel/Vauxhall added abs to the options lists for the Carlton, Senator and Monza. At the same time, Lucas Girling launched their scs (stop control system) as a low-cost option on the Escort, bringing all the advantages down-market.

Essentially, there are three elements in any abs system: wheel-speed sensors, a logic controller (or processing unit), and a hydraulic modulator that reduces (and reinstates) brake pressures. On a typical Bosch system, wheel speeds are measured by electrical inductive sensors positioned close to the front and rear wheels (or by a single sensor at the differential in the first-generation systems used on early rear-

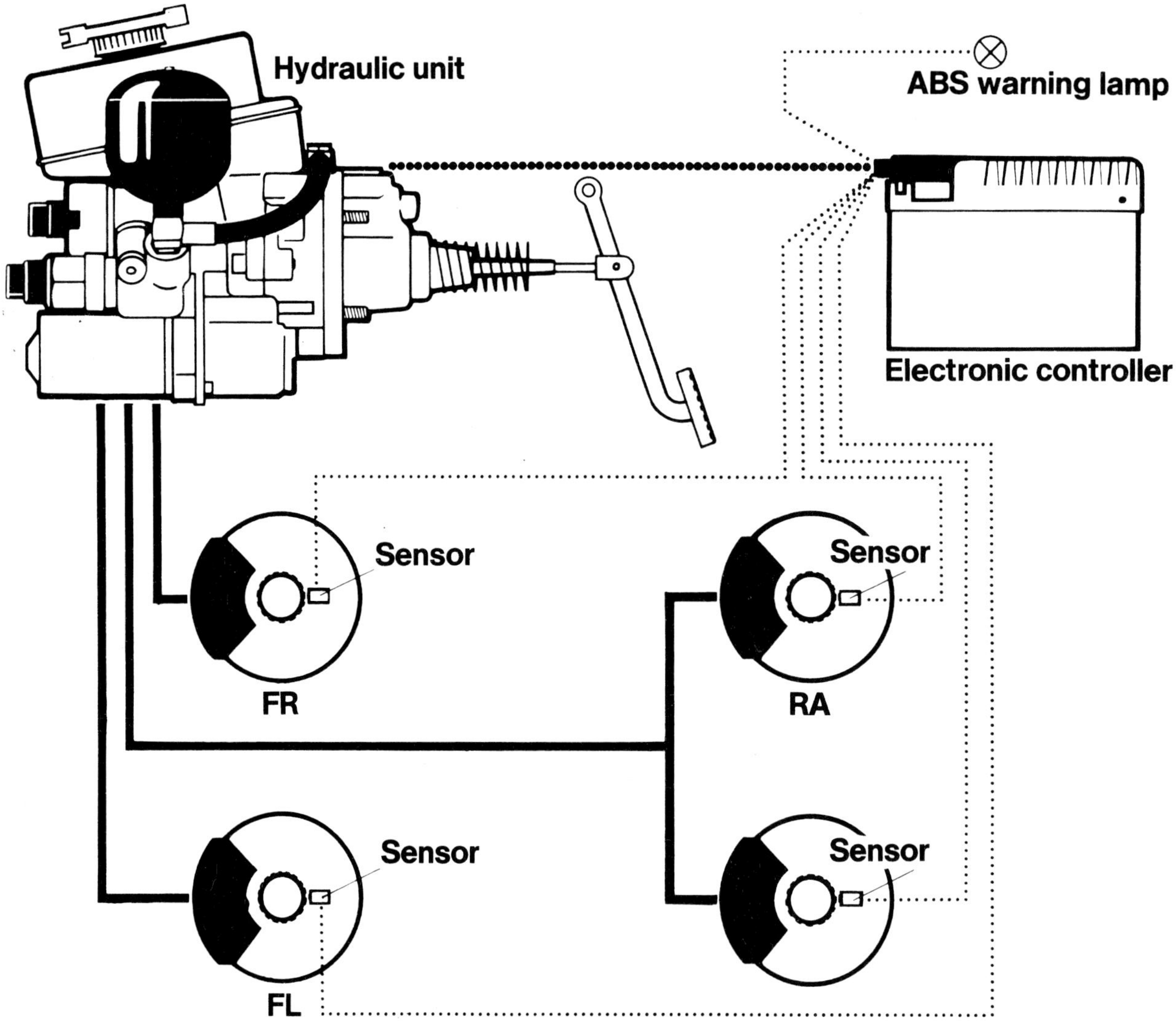

In an abs installation there are usually four wheel sensors feeding speed-change inputs to a microcomputer which controls the pressure supplied to each wheel caliper

wheel-drive applications). The signals from the sensors are fed to an electronic controller, which evaluates their rate of change and their relative characteristics to detect potential skid conditions. It also controls the braking pressure in the wheel cylinders through solenoids that form part of the modulator system.

The hydraulic modulator comprises an accumulator chamber for each brake circuit (to store surplus brake fluid temporarily during cycling), an electric return pump (to move spare fluid around the system), and solenoid valves for each part of the hydraulic system. These valves can connect the wheel cylinders to the master cylinder or to the return pump, or close them off from both.

There are four sensors on the Teves system, and the controller is integrated with the brake booster, saving under-bonnet space and some weight. It uses high-pressure hydraulic fluid, taken from the power-steering pump, to operate the brakes, which gives a more positive and progressive pedal feel under normal applications. When the controller receives wheel-speed signals (processed about 8000 times/sec at 80 mph/128 km/h, for example) that indicate imminent wheel-lock, anti-lock cycles up to 12 times/sec are transmitted to the modulator under a control algorithm that ensures much smoother effects than the Bosch controller.

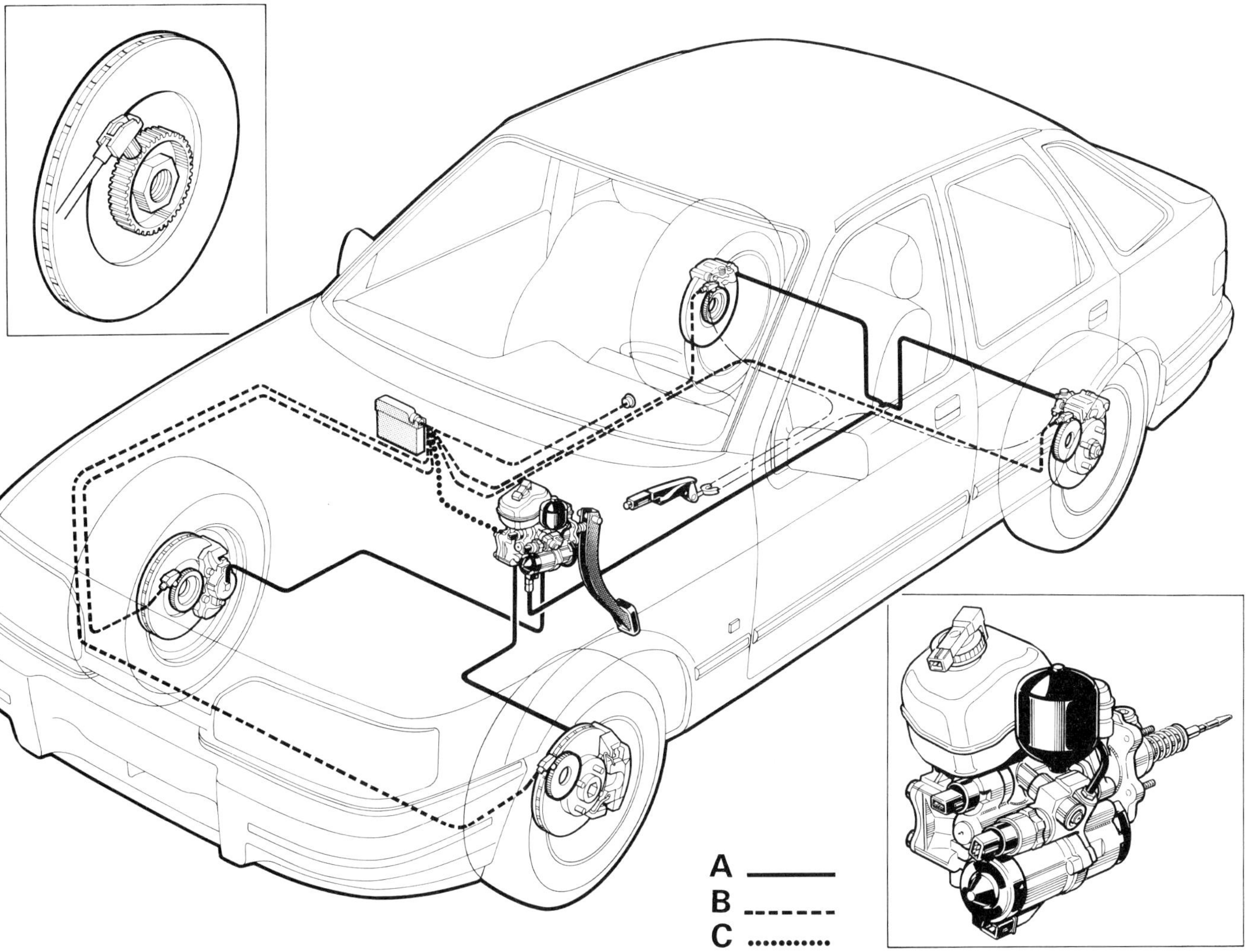

Three different types of circuits are required with abs. Conventional hydraulic brake lines (A) are supplemented by sensor connections (B) and the master control (C). Wheel speed is sensed by inductive pulses (inset left) and hydraulic modulation is performed by a special booster pedal assembly (inset right)

A sophisticated safety concept, adopted by Teves, uses two independent microprocessors to monitor the function of the total system and each other. In normal operation they both generate identical signals, but each can function independently. That logic eliminates the need for start-up test cycles and provides periodic valve monitoring to check operation during normal running.

Ahead of Lucas Girling, Honda lowered abs prices with an anti-lock system for the 1982 Prelude, which was launched at about half the price of an electronic Bosch system. Unfortunately, it failed to make much of an impact. At the time, it was unusual in being back-pressure modulated. That means pressurized hydraulics, at 3000 psi, are used to reduce the braking effect on the wheels, which are braked normally by a conventional, vacuum-boosted system (in which, Honda claim, operating pressures as high as 2500 psi are possible). A microprocessor senses when any wheel deceleration begins to exceed the car's deceleration by more than ten per cent and reduces the pressure to that pair of wheels, preventing locking. It also uses what is described as a high-select front, low-select rear control. This ensures that the front wheels will always apply more braking force than the rear, providing greater stability.

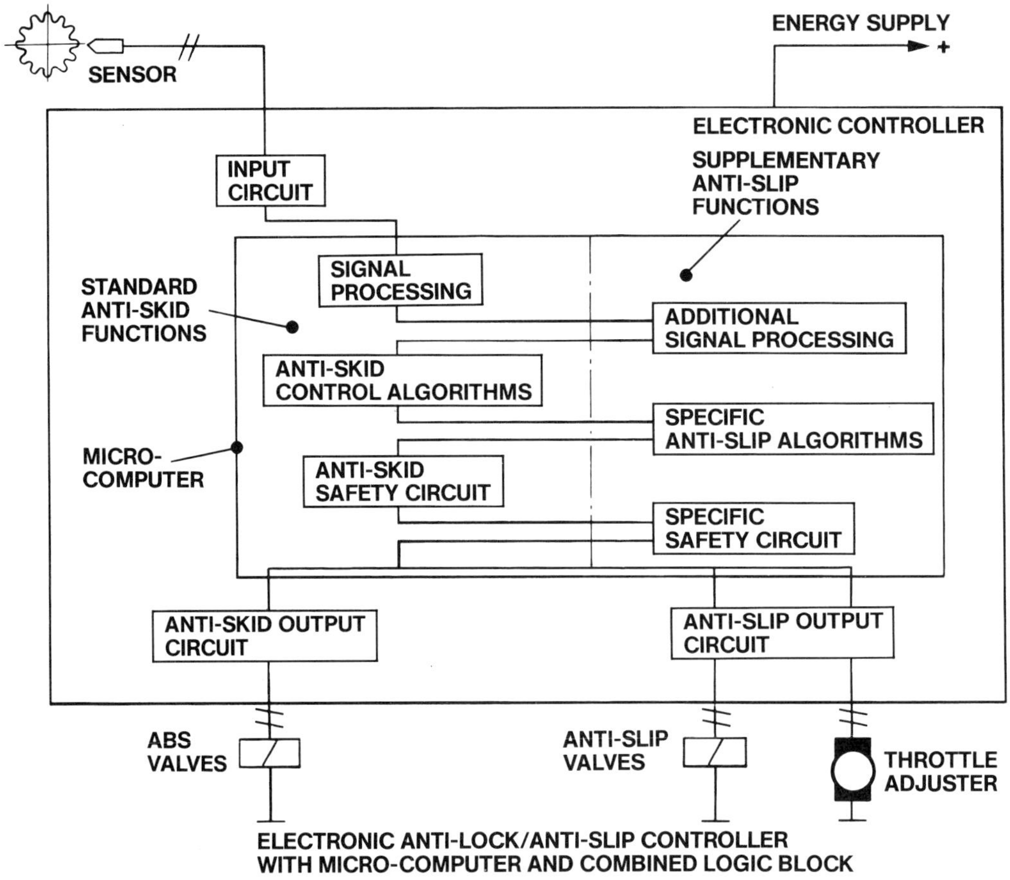

The logic controller on the Teves abs system is suitable for a combined anti-lock/anti-slip version that can prevent wheelspin as well as braking

A different approach was also taken by Lucas Girling when they deliberately turned their backs on electronic abs in 1981. Despite more than 20 years' experience of anti-lock braking equipment (on trains at first and then cars and trucks during the late 1960s and early 1970s) and making the successful transition from analogue to digital control systems that matched Bosch technology, they chose to develop a lower-cost alternative. The viability of expensive, electronic applications for large cars was shrinking with the market, together with any opportunities to recover the investment needed to reach production.

After two successive oil crises and a steady downsizing of the market, it was projected that sales of luxury cars would account for less than ten per cent of European sales by 1985, with abs taking only three to ten per cent of that. Against the risk of being priced out altogether, Lucas Girling looked for another direction of attack. At that time, they were working on an anti-lock system for the front wheels of motorcycles.

To keep costs and complexity under control (and eliminate the difficulty of providing reliable electronics in the harsh, vibrational, unprotected environment of a motor bike), they had adopted a hydro-mechanical approach. It employed the simple expedient of an overrun flywheel to sense wheel-lock, a dump valve to release line pressure and an integral pump to restore braking. The system was proving very successful in preliminary tests, despite the low inertia of a motorcycle wheel assembly.

In parallel with this development, the Girling Brake Division was desperately trying to reduce the cost of abs in order to reach a much larger volume of the new-car market. As a first step, they applied a simplified, two-channel, electronic abs to front-wheel-drive cars with X-split hydraulic systems. This proved that although it was still possible to lock one rear wheel under heavy braking on a surface with a split coefficient of friction (conditions where one side of the car is on tarmac, for example, and the other on ice), stability was affected very little indeed and full steering control was retained at all times through the skid-protected front wheels.

With an eye on the front-wheel-drive market (by now taking around 60 per cent of new-car sales), a customer cost projection of well under £400 ($650), and spurred on by a keen interest from Ford, Girling brought the two elements together into a mechanically-driven, two-channel, non-electronic system for cars. Initially, it was fitted to a 200-vehicle Ford fleet for road trials and service validation in the hands of company personnel, as a follow-up to well over a million miles of joint development testing.

The simplicity of the Lucas Girling system is a measure of its brilliance and its extensive development. Its control logic is based on electronic data and experience gained in earlier development, transferred to a simpler, and inherently more-reliable, hydro-mechanical system.

Basically, there are two elements to each channel of the system, which uses no external source of power, other than pedal pressure and the drive for its modulator. It relies totally on a diagonally-split hydraulic system, fitted universally to front-wheel-drive cars to meet current braking regulations, and the inclusion in those systems of rear-brake apportioning valves which prevent rear wheel-lock under most conditions.

By starting from that basis, it has been found that perfectly-adequate slide protection can be provided by monitoring the angular decelerations of each front wheel only. Load-sensitive rear valves that are slightly more sophisticated than the simple g-valves usually fitted are essential to cover the full range of laden conditions, but otherwise the rear wheels take care of themselves because they can never both lock at once. If one does lock, it will always be the one on the least slippery surface and, therefore, the one with the least effect on stability.

Each front wheel drives a modulator unit via a toothed rubber belt, reinforced with Kevlar and nylon faced, from teeth on the outer housing of each inboard constant-velocity joint. Wheel speed is geared up by a ratio of 2.8:1 by the relative size of the belt sprockets. Mounted on each side of the transmission casing, the modulators are inserted in the hydraulic lines between the master cylinder and the point at which each branch divides between front and rear. The modulator contains an integrated sensor, pump and pressure modulator. Because of the diagonal interconnection, it controls one front wheel and the opposite rear through the apportioning valve.

When the sensor detects a deceleration in flywheel speed greater than about 1.2 g, its inertia causes it to overrun its ball-and-ramp drive and generate an axial movement that moves a lever connected to a dump valve, relieving the line pressure. At the same time, a cut-off valve isolates the master cylinder from the brake, and its residual pressure is used to bring a small pump plunger into contact with an eccentric cam on the driven shaft running through the unit. As the dump valve is still open, no pressure is generated yet.

The pressure reduction in the brake line releases the brake and allows the wheel to accelerate, increasing the input speed to match that of the decelerating flywheel, which is being slowed by the friction in its ball-and-ramp mechanism. When the two are synchronized, the dump valve closes, allowing the pump to develop pressure in the brake line equal to that in the master cylinder. At this point, the system reverts to normal and, if necessary, the cycle repeats itself as often as is needed to bring the car to a halt.

Girling claim that the dump sequence is actually faster than an electronically-controlled system (3 instead of 5 milliseconds), although a full cycle takes about the same time. Cycling time reaches only 3–5 Hz, which, in theory, is slower than an electronic abs, but close to the frequencies used in practice to keep below suspension resonances.

It is only a mater of time before other manufacturers of front-wheel-drive cars follow the Ford lead and offer the same low-cost option, nor will it be long before other brake system producers introduce competitive systems. As customer awareness of abs advantages reaches the lower end of the market, demand is bound to grow.

Driving with abs through a normal UK winter, even in the warmer climate of the south-east, only endorses its value as a safety feature—not just in the limits of braking, which can be extended slightly into corners under slippery conditions, but in dramatically-improved steering control, which is almost unaffected by surface conditions. In Germany, research has proved that about one accident in 13 could have been avoided if steering control was not lost under emergency braking. Although the laws of physics cannot be cheated by any kind of abs, the system does give all drivers much greater margins of control.

A conflict of interests—abs and four-wheel drive

Although abs has come a long way since the first Dunlop Maxaret units appeared, there are some mechanical drive configurations which work against it. The most significant of these are the part-time, four-wheel-drive systems available on an increasing number of popular small cars, and the permanent four-wheel-drive arrangement offered with manually-activated differential locks on the Audi Quattro.

The use of a three-channel system (one for each front wheel and one for the rear axle), introduced to save 25 per cent of the cost of sensors, prevented abs being applied to four-wheel-drive Audis for quite a long time. The problem with 4 × 4 drive systems, as far as abs is concerned, stems from the interconnection between the wheels—the same feature that makes these machines so attractive in traction terms.

In slippery conditions, every car reaches the same physical limit, set by the friction available between the tyres and the road. Excessive braking causes the wheels to lock and the tyres to slide. To generate cornering forces, a tyre must rotate, so a locked front wheel cannot be steered and a locked rear wheel may not run straight. In terms of the total grip available, set by the laws of physics, it doesn't matter in which direction the forces are applied; braking in a bend decreases cornering power, and a spinning drive wheel can easily slide across the road.

For maximum steering control under heavy braking, therefore, front wheels should never reach the point of locking, and for maximum stability under heavy braking, rear wheels should never reach this point either. So in most conditions a car with anti-lock brakes can be made to stop in the same distance as a conventionally-equipped car, and in some conditions it may stop in a shorter

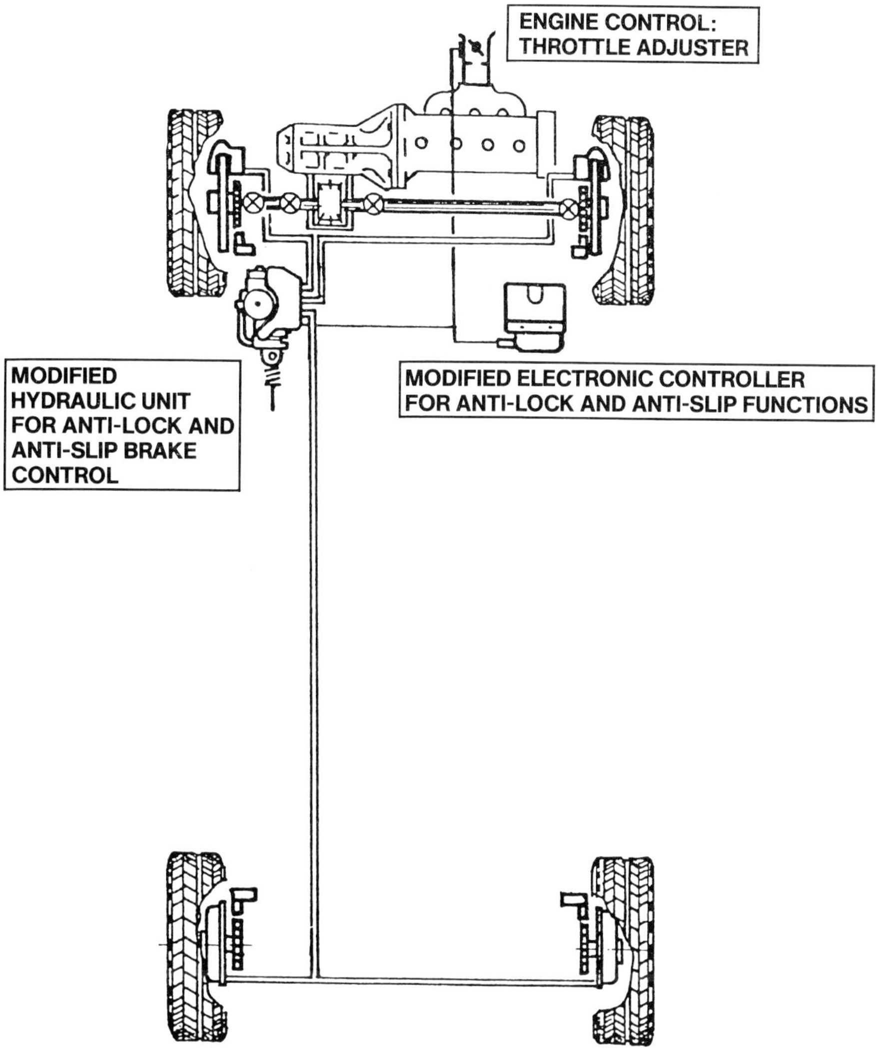

When traction is lost on slippery surfaces, the controls reduce the throttle opening before the spinning wheel is braked to divert the drive to other wheels

distance. On soft, fresh snow, or loose gravel, a locked wheel can collect a wedge of surface material ahead of the tyre and might stop a fraction more quickly, but there will be no steering control or rear-wheel stability.

To allow for these rare and exceptional circumstances, and biased by their good sales record in the traditional European 'snow' markets, like Switzerland, Audi provide a unique abs deactivating switch. All the other manufacturers offering abs provide a consistent level of control and security, taking the onus off the driver.

However, the real conflict of braking interests comes with the rigid central connection for the 4 × 4 system, which is engaged with the differential lock on the Audi Quattro range and used in the part-time layouts offered by Nissan, Subaru, Suzuki, Toyota and all the rest. Current braking legislation, quite rightly, requires that all cars lock their front wheels before the rear, which can prevent maximum retardation levels in some slippery conditions but which generally leads to more-

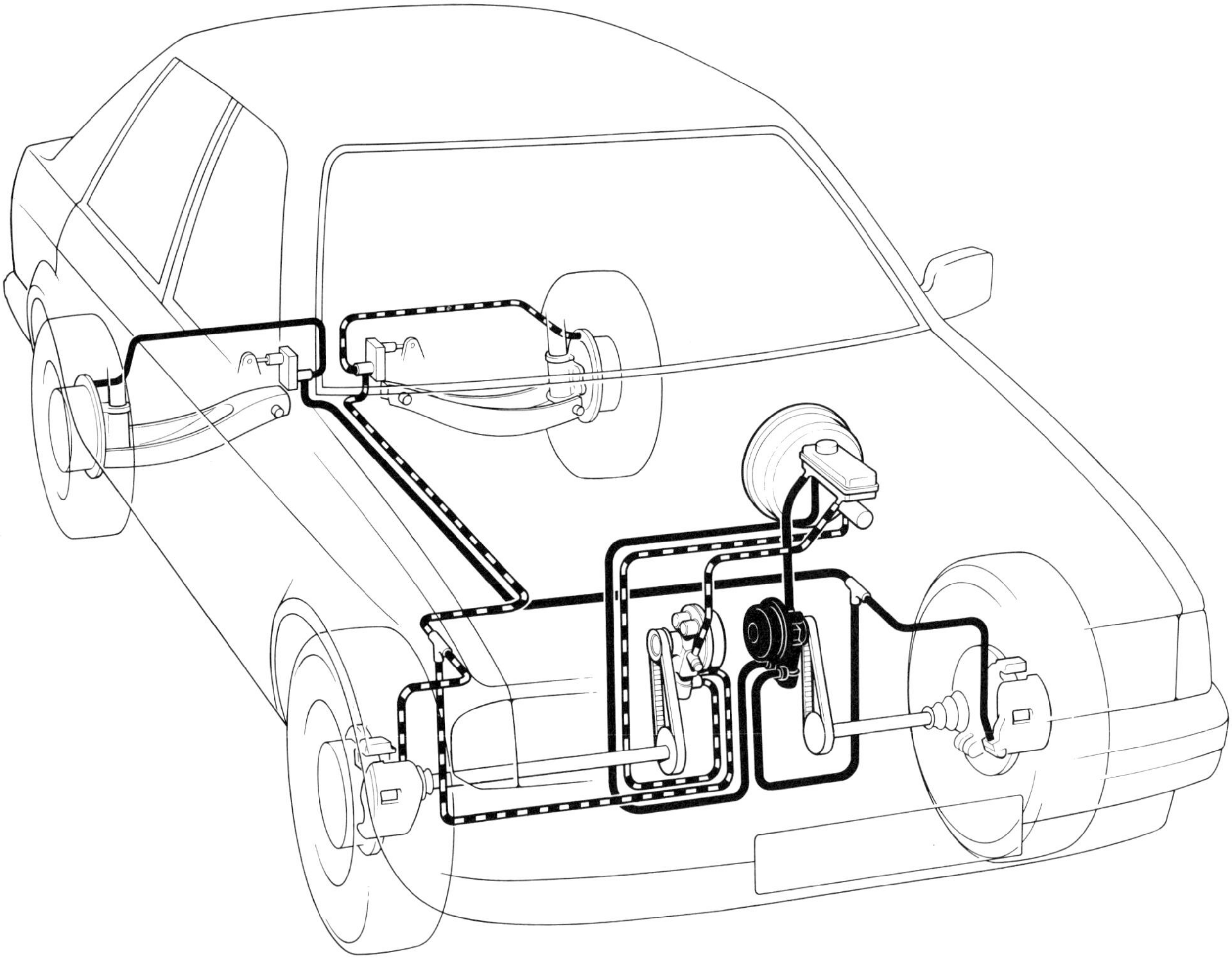

ABOVE *The simplified Lucas Girling stop control system (scs) uses only two modulator units, belt-driven from the front drive shafts. It relies on the diagonally-connected split hydraulic system to give rear-wheel skid protection*

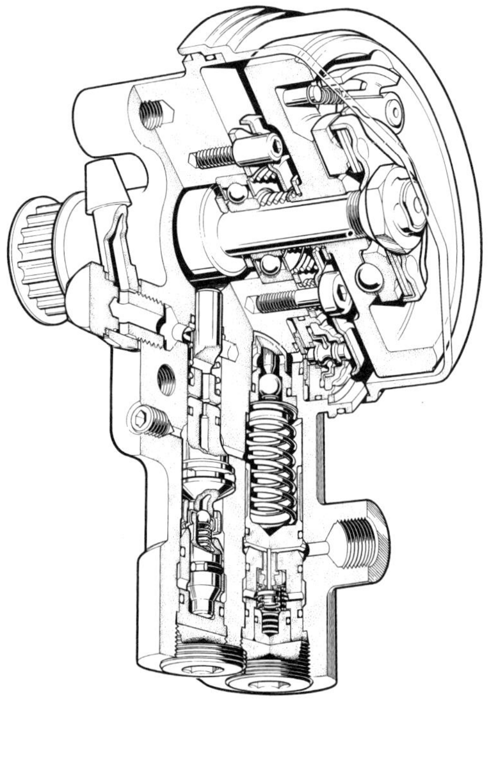

RIGHT *There are no electronics in the Lucas Girling scs modulators, which work on hydro-mechanical principles only. Pressure to the brakes is reduced under the influence of overrun flywheel sensors and bleed valves*

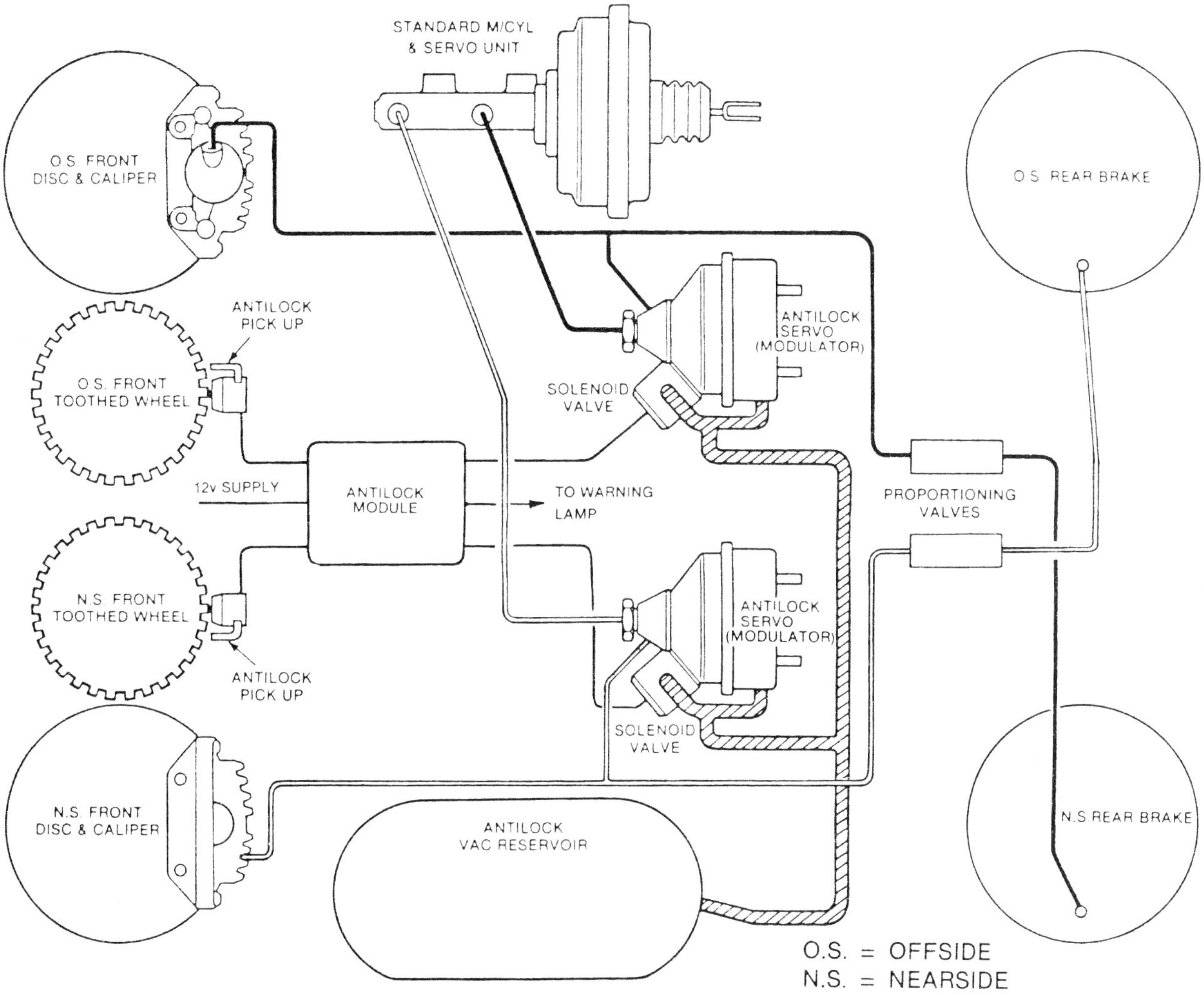

Automotive Products' low-cost abs also uses only two sensors, but it employs electronic signals and solenoid modulators. Both systems are for front-wheel-drive cars only

stable emergency braking. When the drive is split rigidly between front and rear, without a differential to allow different things to happen at each end of the car, both front and rear wheels are forced to turn at the same speed (strictly speaking, the average speed of the front wheels must be the same as that of the rear because the front and rear differentials do allow differing speeds left to right). This raises the threshold of wheel-lock because the fronts are forced to turn by the rears up to the point of rear wheel-lock.

When the wheels do lock, they all lock together, providing the maximum risk of the car spinning. This has been proved quite conclusively in a technical paper prepared by VW research engineers B. Richter and J. Maretzke. It reports on a series of driving manoeuvres carried out on good and slippery surfaces with a full range of mechanical layouts.

On a split-μ surface (differing friction coefficients under left and right wheels), front-wheel-drive, rear-wheel-drive and four-wheel-drive vehicles (the last with a centre differential or viscous coupling) showed the same deviation from a straight line under heavy braking. The rigid, centre-drive 4×4 system, on the other hand, deviated by 50 per cent more—to the point where the authors said, 'the situation should be looked upon as scarcely controllable'.

To design a foolproof abs system that can cope with all the complications of these dynamics has been impossible so far. Therefore, Audi switch it out automatically as soon as the centre differential lock is activated on the first-generation Quattro system, and when the rear differential is locked on the later systems which benefit from a torque-sensitive, limited-slip centre differential. On the Mercedes-Benz 4-Matic cars and VW Golf Synchro, the 4×4 system is disconnected automatically under braking, to solve the problem in a different way.

VW and Audi seem to be in direct conflict with each other when their standard product offerings are compared in this respect. Richter and Maretzke summed up their paper by concluding that the greatest deceleration in straight-line braking is possible with selectable four-wheel drive. They confirm, though, that the advantage is bought at the price of poor directional stability, a fact which can be seen from the directional deviations produced by their braking tests on split-μ surfaces.

Richter and Maretzke also conclude that the majority of drivers would not be capable of controlling a car in these conditions, which are particularly dangerous in the case of a vehicle with its rear differential locked. Braking on a bend also gave a similar result.

13 Future trends

In the field of suspension design, the greatest steps forward over the next few years will result from progress in electronic systems and computer control at levels that were inconceivable a few years ago. Stimulated by advances in computer technology in other fields, the latest high-speed, high-power, 16-bit chips are already appearing on production cars to control exhaust-emission equipment and have plenty of spare processing capacity for other tasks.

A prime example of today's state-of-the-art, on-board computing systems is the Ford EEC IV microprocessor, which is built in the USA by the Ford Electrical and Electronics Division and fitted to several European cars to manage their fuel-injection and ignition systems. It contains a total addressable memory of 64 kilobytes, with 41 input/output lines, and can operate at speeds of up to 16 megahertz (equivalent to making 16 million processing decisions per second). In its current applications, only 32 kilobytes are being used, which leaves plenty of capacity to drive electronic transmission controls and an advanced suspension system. These functions were included in an experimental Ford prototype called ELTEC, which was built in 1985 to show the direction of future developments.

ELTEC used electronically-controlled, air-suspension units to maintain a constant ride height and consistent ride qualities, regardless of load, automatic adjustments being made to the pressure inside the air springs. The basic setting was particularly soft and comfortable, while a firmer ride was selected automatically for high-speed driving.

Nissan went even further with their CUE-X design study, using electronic air suspension to control spring rates, ride height and attitude control as well as variable damping, all combined with four-wheel drive, anti-lock brakes and a dual-mode, four-wheel steering system. Electrical signals representing vehicle height, road speed, steering input, braking forces, throttle position and transmission selector position were fed into a microprocessor, which operated compressors and valves to inflate or deflate their air springs according to algorithms devised during extensive development of test vehicles.

CUE-X offered two different spring rates (a simple, stepped control of stiffness) and three damping settings (again, as stepped functions) combined with three different ride-height values. Various permutations were offered by the system, which automatically overrode the driver's choice from two settings (normal or sporty). The natural progression from such a system would be to a variable and adaptive 'smart' control that learned from the experience it gained during actual operation. Such levels of artificial intelligence are already well advanced and likely to appear before the end of the 1980s.

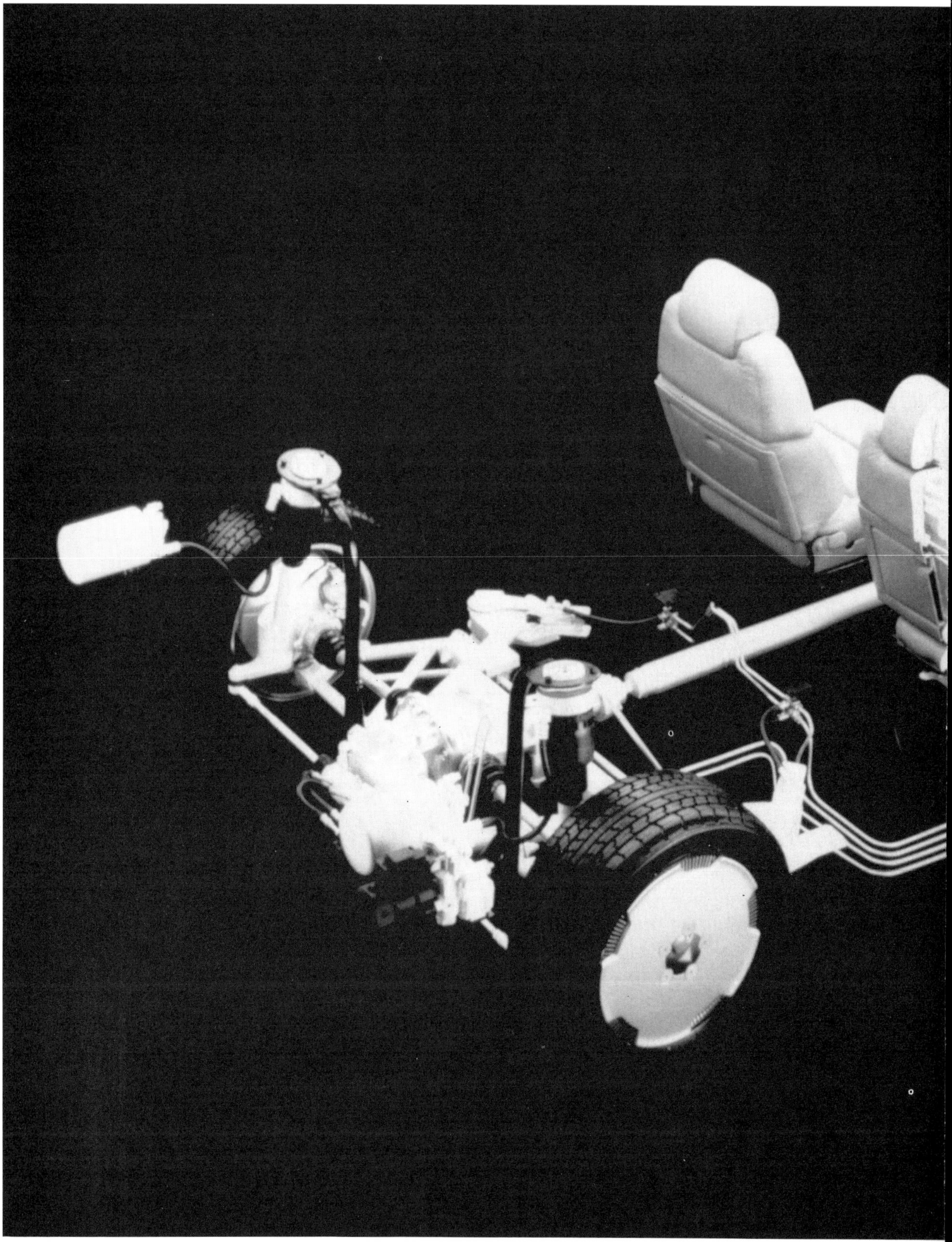

Probably the most advanced design in suspension technology to date, the Nissan CUE-X uses on-board computers to provide a wide range of spring and damper settings. Electronic sensors supply running data from which algorithmic outputs control fast-acting, variable valves

Ford's advanced suspension concepts were built into a design study called ELTEC, first shown in 1985. It uses their powerful EEC IV 16-bit microprocessor to manage an advanced adaptive suspension in addition to a 'drive-by-wire' engine system

In the parallel field of steering control, a race is already under way between ZF in Germany and General Motors and Chrysler in the United States to produce the first electronic power steering, frequently referred to as a 'steer-by-wire' system. The term refers to power steering where the torsional input at the wheel is monitored electronically and used to control an electric servo motor on the steering rack. There is still a positive, mechanical linkage between the steering wheel and the front wheels, but the system does away with hydraulic pumps, valves, pipes and rams, reducing the components required by 85 per cent and the weight of the system by up to 7.7 lb (3.5 kg). It is logical, versatile and much simpler than a conventional system.

The General Motors system was conceived by their Research Laboratories in 1979 and has been under development by the Saginaw Steering Gear Division since 1982. It will probably appear first on the Pontiac Fiero, where the mid-engine location complicates the installation of an engine-driven hydraulic pump and has led to a costly, hybrid, hydraulic/electronic system. According to GM, electric power steering (eps) is more compact than conventional hydraulic systems, absorbs less engine power (as it only needs energy when it is working) and hence improves fuel economy, and still operates after the engine has stalled.

From the functional and quality points of view, there are two more worthwhile gains provided by eps. One is that it would eliminate many of the customer complaints derived from hydraulic, power-steering systems, which tend to be noisy (especially at full lock when the blow-off valves are operating) and prone to leaks. Therefore, quality ratings would improve by a change to electrical power.

The second gain is that the degree of assistance could be programmed more precisely to suit such variables as engine weight, tyre equipment, driver preference and speed. An eps arrangement also opens the door to the development of four-wheel steering and integrated suspension, braking and transmission systems, all controlled by an advanced, on-board microprocessor. The latest active rear-wheel steering controls, like the Nissan HICAS system, with less than a degree of angular change, could actually be driven electrically to reduce their cost and complexity significantly.

GM's eps is based on a manual rack-and-pinion with an electric motor attached to one side of the pinion housing and applying assistance through gears to the lower end of the pinion shaft. It monitors driver input through an almost conventional, torsion-bar sensor inside the column. The signal from the sensor is converted from analogue to digital and processed to energize the windings of the motor using pulse-

The latest new model to feature advanced electronically-controlled suspension, abs and speed-sensitive power steering is the Opel Senator. It sets a technological pattern for all new luxury-car designs

The ZF Servotronic power steering used on the Senator varies the level of assistance to give correspondingly light effort for parking and heavier loads at speed

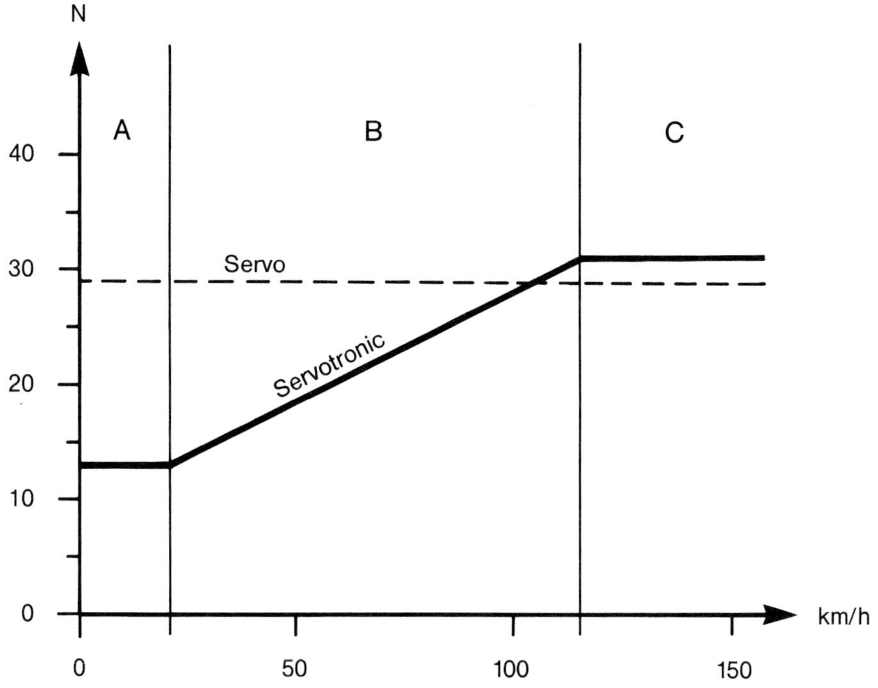

Servotronic provides more power assistance (A) at low speed by increasing the pressure (B) to the hydraulic servo inversely with speed (C)

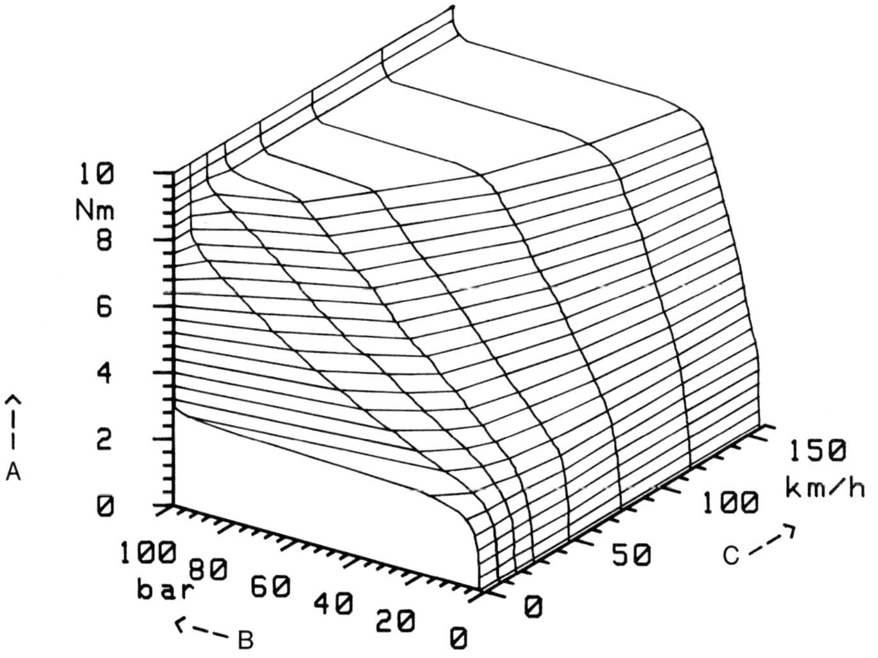

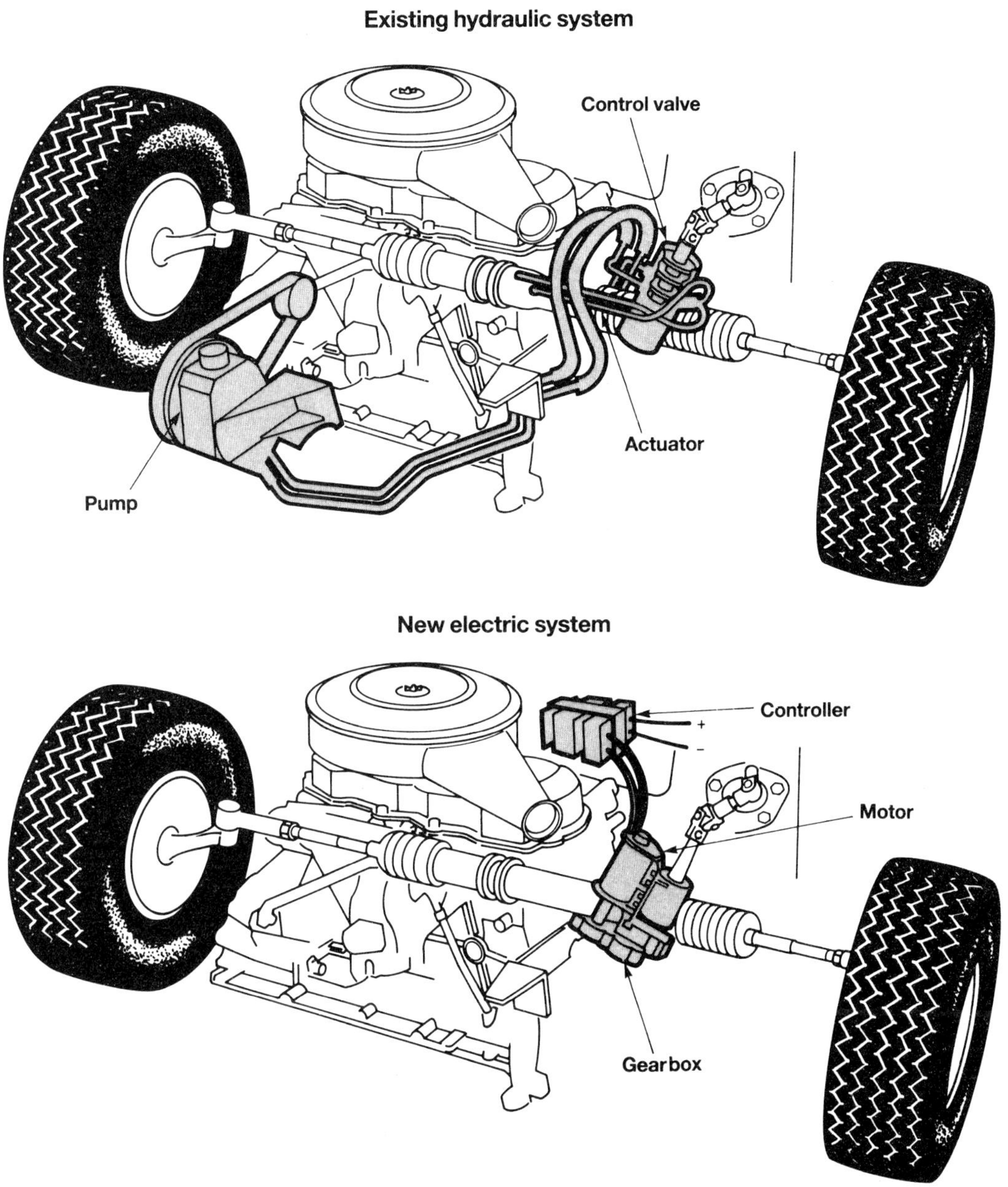

Saginaw speed-sensitive steering is electrically operated, eliminating all the complicated hydraulics of conventional power steering. The system could easily be adapted for rear-wheel steering control

width, modulated voltage and battery power from the vehicle's electrical system.

Like the Teves anti-lock braking system used by Ford and VW, GM's eps has two microprocessors to check on each other and provide back-up if one fails. The system also shuts down automatically if there is an electrical failure.

However, the main advantage of eps, from the dynamic angle, is that it can be programmed easily to be speed sensitive, like the power steering introduced 16 years ago on the Citroën SM and the new ZF Servotronic power steering on the BMW 7-series and Opel Omega range. It is also quite feasible to detect front-tyre slip angles from the feedback into the system and vary the power from the engine as a traction aid in corners, or to modulate the level of assistance to improve driver feel in changing road conditions.

TRW Inc., steering suppliers to Chrysler and other non-GM producers (and

Bilstein has produced this fast-acting electronically-controlled adjustable damper for the advanced Porsche 959 largely using aluminium, to reduce the unsprung weight and improve heat dissipation

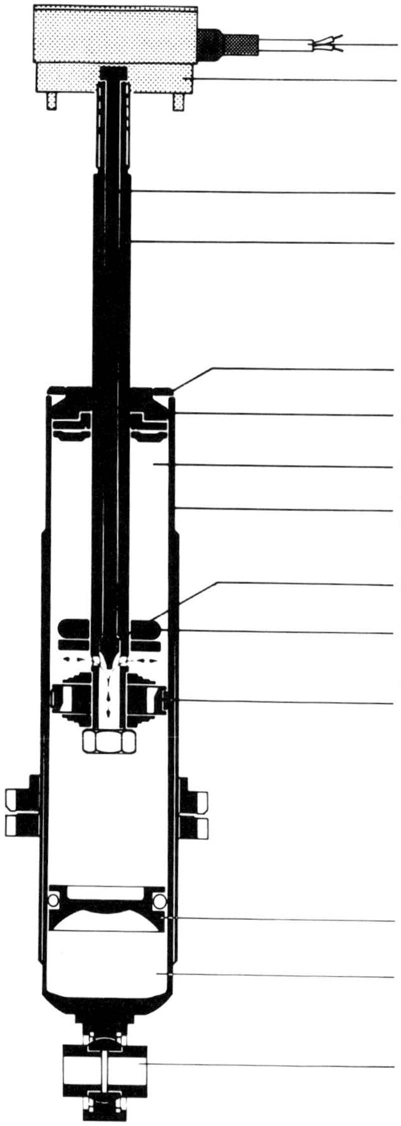

parent of the UK-based Cam Gears Ltd), claim to be two years ahead of the competition with their Powertronic electrically-controlled steering system (ecs). This has an armature wound around the central part of the steering rack with a ball-nut assembly that converts the armature's rotary motion into a reversible, linear thrust acting directly on the rack teeth. With the torque sensor on the pinion input shaft, the entire system is integrated on the rack assembly, which further reduces weight and complexity.

TRW are optimistic about the future of their system, in which they have invested about £10 million ($16 million) so far. They claim that they will start production before the end of 1987 and produce about 30,000 ecs units in the first year, rising to 500,000 by 1990. By 1992, it is expected that more than 3 million electronic power-steering systems will be produced in North America per year.

Cam Gears are testing ecs for future use by European producers, and ZF are working hard on a competitive arrangement. In the meantime, the German suppliers have the speed-sensitive Servotronic system. It uses a speed-modulated,

EAS System Illustration Showing Location of Key System Components

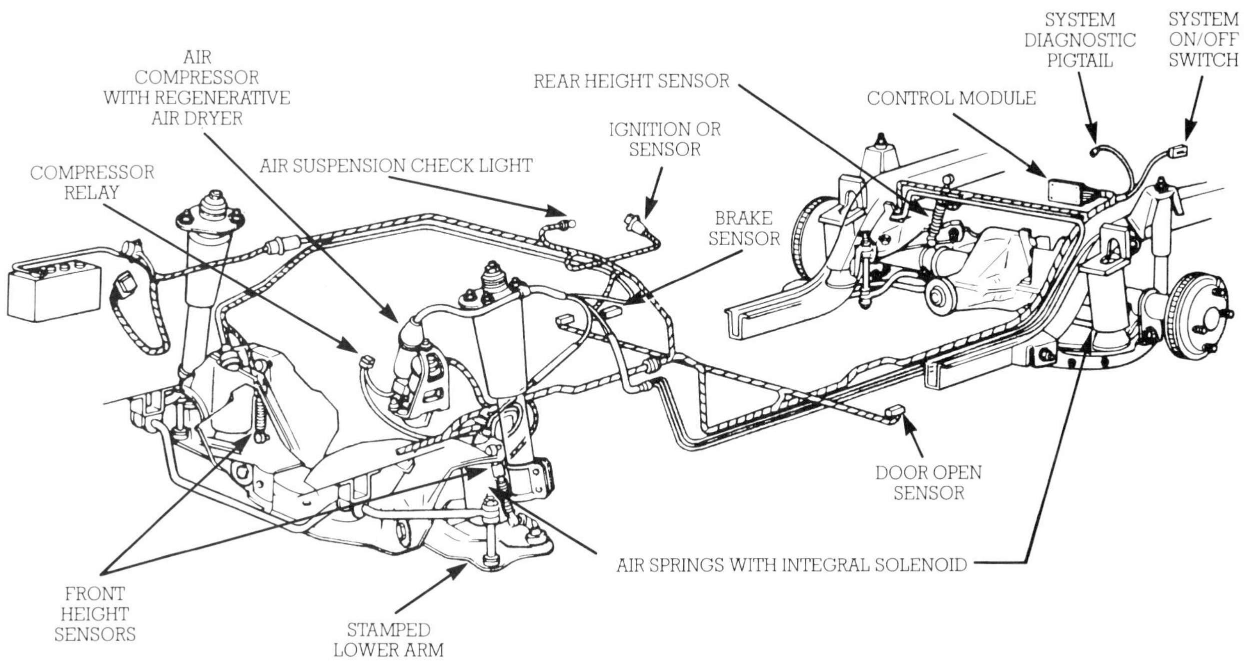

ABOVE *Cars of the future will incorporate sophisticated computer-controlled systems like this Ford US design already offered on some top-of-the-range models*

RIGHT *System logic for advanced suspension controls must detect several different sets of operating conditions. On this Ford design, brake and door position sensors are used to provide intelligence on the state of the vehicle*

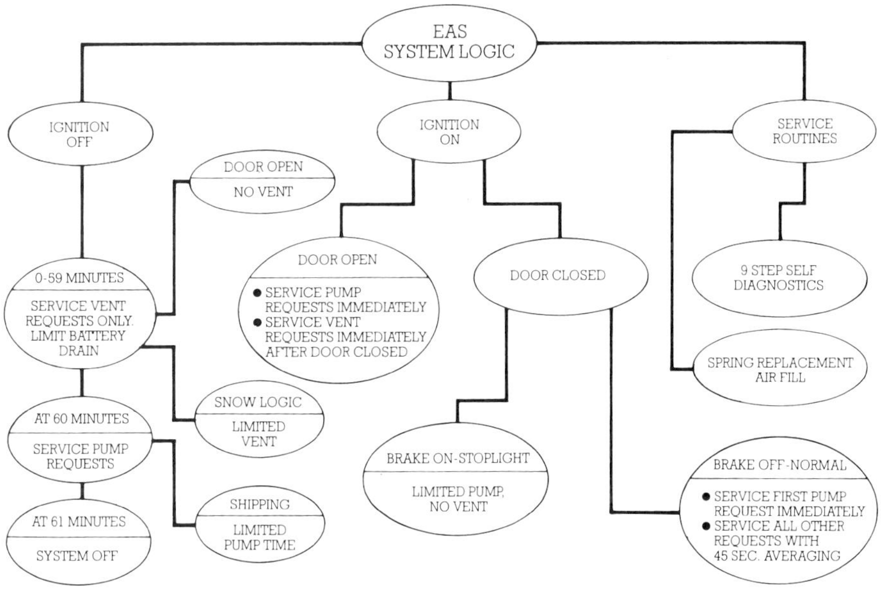

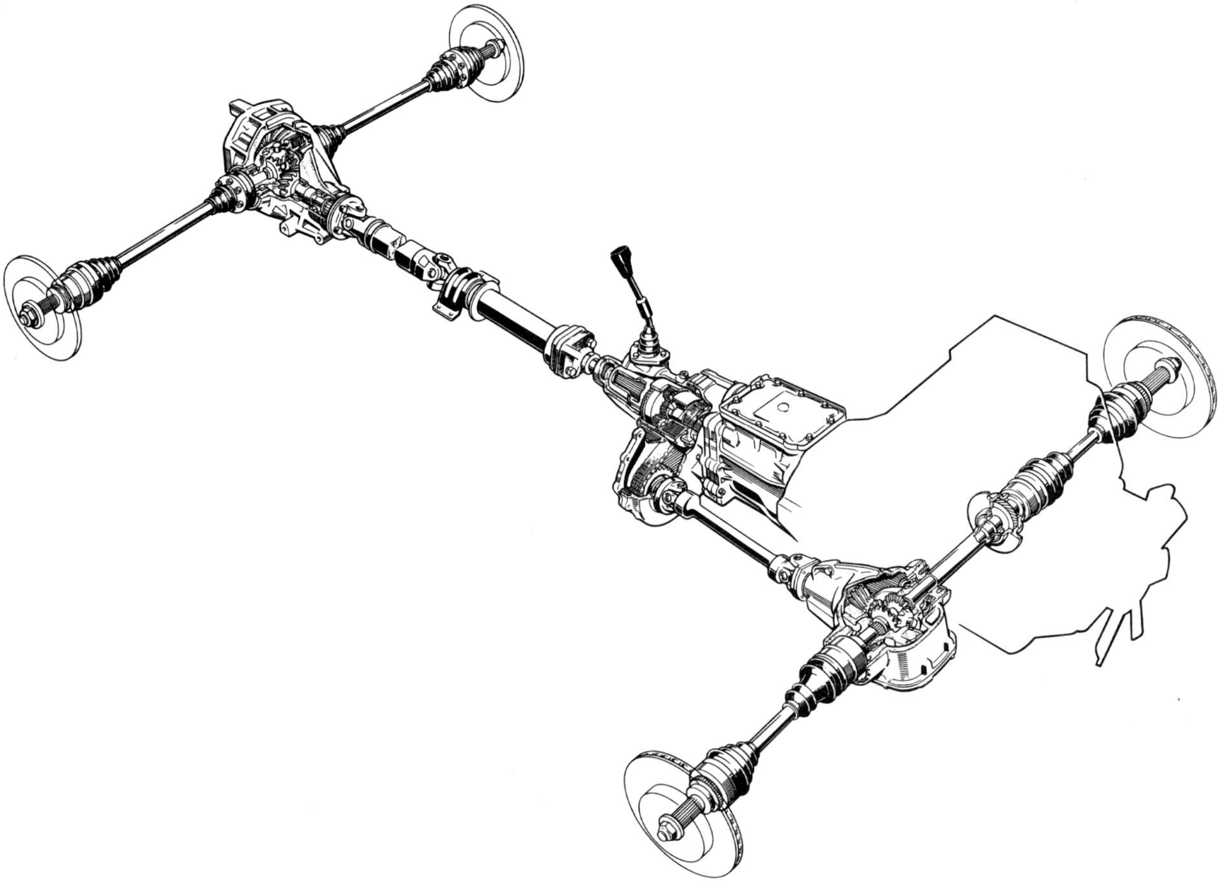

ABOVE *There seems little doubt that simply-designed four-wheel-drive systems will become much more commonplace by the end of the decade, combined in many cases with abs and rear-wheel steering*

RIGHT *For computer control of suspension dynamics, the car of the future will be able to sense tractive effort (1), braking (2) and lateral forces (3), vertical reactions (4), yaw moments (5) and moments of inertia (6), all within a few milliseconds*

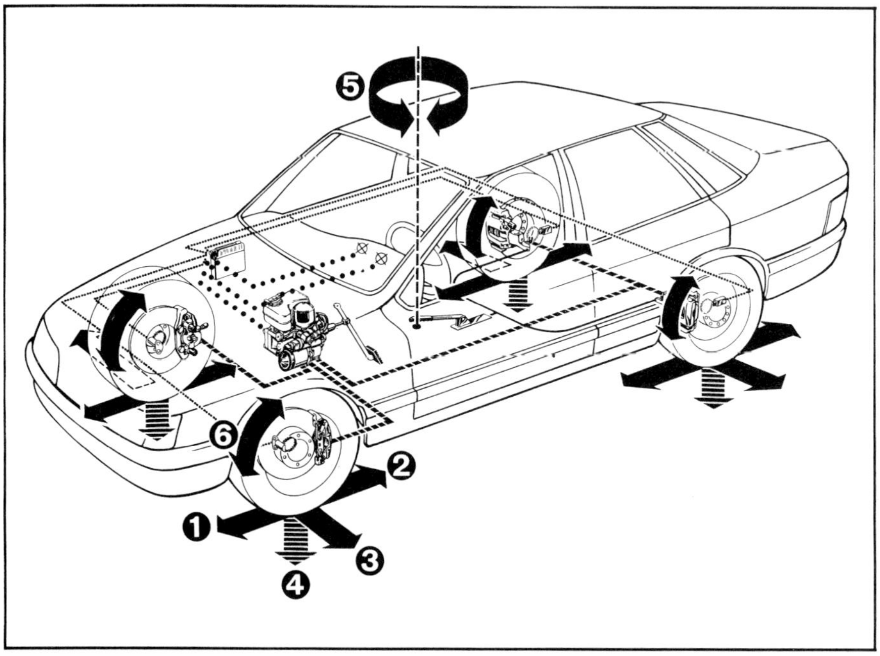

hydraulic input to vary the level of power assistance progressively, in accordance with a two-stage, linear decay curve, between speeds of 12 mph (19.3 km/h) and 50 mph (80 km/h), and at a reduced rate between 50 and 75 mph (120 km/h).

As on the original Citroën SM power steering, the theory behind the Servotronic system is that light steering at high speed makes a car over-sensitive and the driver nervous. The compromise provided by fixed-level assistance is either too heavy for parking or too delicate at speed.

On the SM, there was so much assistance at rest that the wheels would power themselves to the straight-ahead position hands off, and as there were less than two turns lock-to-lock, it was difficult for drivers new to the car to handle the system smoothly. After a few hours, though, the SM was like a huge sports car, with similar responses to those of the Lotus Elan, which had steering geared to the same ratio.

Luxury cars, such as the larger Mercedes-Benz models, Jaguars and even Rolls-Royces, would be much more nimble with variable electric power steering, and generally far more comfortable and controllable with active-ride control, making future prospects in the car industry look very good indeed.

Index

abs 80, 164, 165, 167, 168, 169, 170, 171, 173
AC Ace 25
Ackermann effect 31, 91
Active ride 125, 127, 135, 187
ADAMS computer programme 88
Adaptive suspension 76, 127, 128, 130, 131, 134, 138, 140, 143, 175
Ader, Clément 105
Adler 61
Aerodynamic
  downforce 134
  lift 71
Aerodynamics 65, 105
Aerofoils 134
Air springs 28, 131, 135, 138, 175
Airflow 121
All-wheel drive 173
Aluminium 11, 74, 118
Alvis 61
Amdahl 87
American Underslung 11
American Victor 11
Anechoic chamber 151
Anti-dive 31, 50, 91
Anti-lock brakes 164, 165, 170, 175
Anti-roll bar 27, 46, 50, 55, 58, 130
Anti-squat 31, 50, 51
Aquaplaning 99, 100
Aramids 97
Artificial intelligence 54, 175
Artz press 11
Aston Martin 11
Audi 61, 65, 66, 67, 73, 74, 75, 76, 102, 153, 164, 165, 170, 171, 174
  100 66, 153
  200 Turbo 73
  Quattro 73, 75
Austin 3-litre 128
  Allegro 128, 130
  Maestro 130
  Seven 73
Austin Rover Group 65, 76
Auto Union 109, 112, 115
Automobile Club de France 108
Automotive Products 131
Avon Rubber Ltd 101

Bantam 73
Belly Flopper 73
Berlin Motor Show 17
Birmingham Motor Show 153
BL Cars 65
BMW 7-series 183
Boat tail 14
Body-inertia forces 128, 135
Body roll 31, 46, 49, 118, 128, 139, 154
Bollée, Amédée (et fils) 105
Bosch 164, 165, 166, 167, 168
Bounce 27, 29, 39, 42
  frequency 128
Bouton, Georges 105
Brakes 67, 72, 80, 103, 108, 135, 164, 165, 170, 175
Brasier 108
BRM 113
Broadley, Eric 115
Brooklands 113
Budd 11, 14, 16
Bugatti 109
Bump-induced harshness 158
Burroughs 87

cad 82, 86, 91, 92
Cadence braking 164
Cadillac 131
cam 82, 91, 92, 169, 184
Cam Gears Ltd 184
Camber angle 29, 30, 46, 49, 50, 51, 52, 54, 55, 58, 91, 109, 112, 115, 118, 125, 131, 155, 157
Carcass, tyre 34, 93, 95, 102, 103
Carrosserie Henri Labourdette 14
Castor angle 29
Centre of gravity 11, 39, 40, 41, 42, 127, 129, 130
Centrepoint steering 30
Chapman, Colin 115, 118
Chapman strut 115
Chevrolet Blazer 74
  Corvair 50
  Corvette 25
Christie, J. Walter 61
Chrysler 180, 183
cie 82
cim 92
Circle of Forces 154
Citroën 128, 130, 131, 183, 187
  2CV 128, 130
  Big Six 130
  CX 67
  DS19 130, 131
  SM 183, 187
Citroën, André 14, 61
Coach-building 9
Coefficient of friction 54, 70, 164
Coil springs 25, 28, 34, 35, 36, 45, 49, 55, 109, 112, 121, 135
Collins, Peter 113
Comparative stressing 85
Compliance 28, 29, 33, 44, 50, 51, 52, 53, 55, 60, 88, 151, 157, 159
COMPOSE computer programme 90
Compound beam axle 58
  spring 35, 39
Computer 28, 37, 39, 53, 58, 75, 80, 82, 85, 86, 87, 88, 89, 90, 91, 92, 99, 125, 126, 135, 138, 140, 144, 148, 152, 175
  aids 28, 82, 86
  controlled drafting machines 86
  graphics 82, 87, 91
  integrated manufacturing (cim) 92
Computervision 87
Conjugate points 41, 42
Contact patch 28, 54, 95, 155
Continental tyres 102, 113, 135
Control Data computers 90
Cooper racing car 113, 114, 115, 118
Cooper-Climax 115
Cord cars 61

Cornering 28, 29, 30, 31, 46, 49, 50, 52, 53, 54, 70, 90, 93, 95, 101, 115, 125, 126, 130, 131, 134, 135, 139, 154, 155, 157, 158, 159, 163, 170
 force 29, 93, 155, 159
Crab car, The 75, 153
Cranfield Institute of Technology 125, 134
Crash energy 19
 protection 67
Cray computers 87, 90
Critical damping 38
Crosswind stability 90
CTS tyres 103, 104
CUE-X (*see* Nissan)
Cugnot steam carriage 61

Daihatsu Fourtrak 74
Daimler 11, 19, 109
Damper carding 147
Dampers 27, 28, 34, 38, 39, 45, 47, 49, 50, 53, 54, 55, 58, 60, 72, 108, 121, 125, 129, 134, 135, 138, 139, 141, 143, 144, 145, 147, 148
 single-tube 145
 twin-tube 145
Damping effects 28, 38
Datapoint computers 87
de Carbon 60
de Dion 50, 105, 108, 112
 axle 50, 108, 112
Dearborn 86, 87, 135
Deauville car 105
DEC computers 87
Degrees of freedom 28, 39, 90
Detroit race circuit 125
Digital control 134, 168
 signals 140
Digitizing tablets 85
Discomfort levels 43, 150
Dixon, Freddie 153
DKW 61
Dodge Ramcharger 74
Donington Park race circuit 113
Double wishbones 45, 47, 49, 121
Drag 25, 54, 65, 66, 67, 77, 115, 121, 158
DRICOM computer programme 90
Dropped axles 11
Dual-rate springs 130
Dunlop 93, 99, 102, 103, 128, 164, 170
 Denovo 102, 103
 Maxaret 164, 170
Dunlop, William 93
Durability 19, 55, 82, 88, 91, 128, 144, 145, 147, 148
Dynamic weight transfer 41, 51, 70, 125, 130, 159

Elastomers 97
Electric power steering (eps) 180, 181, 183
Electronic air suspension 175
Electronically-controlled suspension 125
ELTEC (*see* Ford)
Endurance testing 148
ERA cars 19, 75, 109

Failure modes and effect 91
Fatigue-decreased proficiency 151
fe 82, 85, 86
Ferguson 75, 76, 77, 80, 81
 viscous coupling 76
Ferrari Testarossa 76
Fiat 45, 62, 74
 Campagnola 74
 Panda 4 × 4 45
Fibre-optic 86
Finite elements (fe) 58, 82, 85, 86
Firestone 131
Flexibility 19, 26, 28, 34, 35, 47, 51, 52, 88, 92, 157, 158
Ford 17, 31, 44, 45, 46, 52, 55, 62, 63, 65, 73, 74, 75, 76, 86, 87, 88, 104, 131, 135, 139, 140, 145, 148, 158, 164, 165, 169, 170, 171, 175, 183
 105 Anglia 62
 Bronco 74
 Capri 45
 Cardinal 62
 Cortina 17, 19
 EEC IV microprocessor 175
 ELTEC 175
 Escort 140, 164, 165
 Fiesta 55, 62, 67, 145
 Granada 52, 131, 135, 140, 158, 164, 165, 183
 Granada S-Pack 158
 Model T 73
 RS Cosworth 44
 Scorpio 145
 Sierra 52, 63, 158, 183
 Special Vehicle Engineering 44
 Taunus 12M 62, 131
Four-link rear suspension 118
Four-wheel drive 73, 74, 75, 76, 80, 164, 170, 173, 174, 175
Four-wheel steer 153
4 × 4 45, 75, 76, 77, 80, 81, 170, 171, 173, 174
Frankfurt Motor Show 75, 99
Frequency analysis 144
FRESH computer programme 88, 90, 164
Friction 25, 27, 38, 45, 54, 70, 75, 76, 95, 108, 144, 145, 154, 155, 164, 170, 173
 coefficient 95, 173
 dampers 108
Front suspension 14, 25, 41, 46, 53, 61, 90, 91, 105, 109, 112, 121, 130
Frontal area 66, 112, 121
fwd 55, 61, 62, 63, 65, 66, 67, 70, 71, 72, 73, 77, 90, 173

g-valves 169
Gabriel 109
Gas 36, 60, 131, 135, 145
Gas-filled dampers 60
General Automation computers 87
General Motors 17, 87, 92, 125, 134, 180
Geometry, suspension 25, 27, 28, 29, 31, 33, 42, 45, 47, 50, 51, 52, 53, 58, 67, 90, 112, 118, 121, 126, 128, 152, 154, 155, 163
Girling 165, 167, 168, 169, 170
Girling Brake Division 169
Goodwood race circuit 113
Goodyear 100, 101
Gregoire, J. A. 61
Gregory, Ben 61
Greyhound buses 131
Group B rally car 81

Hall-effect height sensors 135
Handling 25, 28, 29, 31, 33, 39, 44, 46, 52, 53, 55, 58, 76, 81, 87, 88, 89, 90, 93, 95, 101, 103, 108, 112, 127, 128, 130, 135, 139, 145, 148, 152, 154, 155, 158
Hanover University 104
Hawthorn, Mike 113
HICAS (*see* Nissan)
Honda 65, 154, 164, 167
 Prelude 164, 167
Honeywell computers 90
Horizontal couple 157
 forces 127
Human susceptibility 19, 151
Hydragas 128
Hydraulic actuators 125, 134
 brake modulator 166
 suspension 125
Hydrolastic suspension 36, 128, 129, 130, 131
Hydro-pneumatic suspension 130

IBM computers 87
Independent front suspension 25, 61, 105
 rear suspension (irs) 49, 70, 154, 157
 trailing arm 31
Indianapolis raceway 61, 118
Insulating bushes 53, 128
Integral steel body 14, 17
Interconnection 41, 127, 130, 169, 170
International Scout 74
International Standards Organization 151
irs 49, 70, 154, 157
Issigonis, Alec 62, 128

Jaguar 49, 51, 99
XJ6 49
Jeep 73, 74
Jensen 75, 164
JJD wheel concept 99
Jounce 31, 91, 140, 141, 148
Juhan, Jerry 99

Kevlar 97, 169
Kinematics 25, 26, 27, 28, 29, 53, 54, 55, 88, 89, 105, 128
Knyff, René de 14

Labourdette, Henri 14
Lada Niva 74
Lancia 14, 62, 67
Lambda 14
Y10 67
Land Rover 74
Langen AG 131
Laws of motion 28
Leaf springs 11, 19, 25, 35, 36, 45, 105, 108, 109, 121, 154
Lever-arm hydraulic dampers 49
Limit of adhesion 155
Limited-slip differential 81
Lincoln Continental 135
Linear-rate spring 35
Live rear axle 49
Load-levelling 29
Lola 114, 115, 118
Lotus 100, 114, 115, 118, 125, 134, 187
Elite 115
Esprit 100, 134
T92 134
Low-profile tyres 55, 145
Lucas Girling 165, 167, 168, 169
scs 165

MacPherson struts 31, 44, 45, 46, 47, 115
map 92
Maretzke, J. 173, 174
Mazda 75, 153
MX-03 75
McDonnell Douglas 92
Mechanical Dynamics Inc. 88
Mechanical trail 29
Mechanics 26, 27
Medical effects 151
Mercedes-Benz 11, 50, 63, 75, 77, 81, 109, 112, 130, 131, 157, 164, 165, 171, 174, 187
220S 131
300SL 157
600 131
S-class 131, 164
V8 165
W25 109
W125 112
W163 112
4-Matic 77, 80
MG EX-E 75, 76
Michelin 93, 99
Microwave transmissions 86
Miller, Harry 61
Mini 61, 62, 67, 102, 128, 129
Metro 128
Moments of inertia 85
Momentum 28
Monaco race circuit 125
Monocoque structures 121
Morgan 45
Morris 16, 128, 129, 130, 131
1100 128, 129, 130, 131
1800 128
Mors racing car 108
Moss, Stirling 113
Motion sickness 151
Moulton, Alex 128

Natural frequency 27, 29, 37, 38, 39, 42
Negative offset 30, 54
Nibel, Hans 109
Nissan CUE-X 175
Bluebird 154, 159
HICAS 81, 158, 159, 163, 181
MID-4 75, 76, 154, 159
Patrol 74
Prairie 75
No-roll suspension 131
Noise, vibration and harshness 46, 60, 91, 150
Norton 113
Numerical control (nc) 86, 87
nvh 91, 150

Opel 17, 53, 55, 66, 87, 90, 158, 164, 165, 171, 183
Ascona 63
Monza 165
Olympia 17
Omega 53, 66, 164, 183
Senator 55, 158, 164, 165
Oversteer 70 109, 112, 155, 157, 163
Owen Organization 113

Package efficiency 25, 67
Panhard 62, 108
rod 49, 58, 112
Panhard-Levassor 14
Paris–Bordeaux–Paris road race 108
Periodic time 37
Pierce-Arrow 11
Pitch 28, 29, 39, 40, 41, 42, 43, 51, 89, 125, 128, 129, 130, 131, 134, 135, 139, 151, 155, 164
frequency 40, 128, 130, 164
Pitching moments 40, 129, 131
motion 39, 41, 139
Plastics 97
Polymerization 95
Polymers 95, 97
Pontiac Fiero 180
Porsche 50, 75, 77, 101, 104, 109, 112, 115, 154, 157, 158, 164, 165, 171
928 101, 154, 158
959 75
R & D Centre 104
Weissach axle 154, 158
Porsche, Ferdinand 109, 112
Power steering 67, 159, 163, 166, 180, 181, 183, 184, 187
Powertronic steering 184
Pressed Steel Company 16
Prime computers 87, 97
Progressive-rate springs 152

Racing-car suspension 105, 114, 115, 121
Range Rover 73
Rear-wheel steering 25, 53, 153, 154, 159, 163
Rebound 31, 38, 58, 91, 118, 125, 129, 140, 141, 148, 157
Renault 50, 55, 62, 66
Dauphine 50, 157
5 66
25 55, 66
Response steer 55
Richter, B. 173, 174
Ride 28, 29, 38, 41, 44, 55, 60, 67, 70, 87, 88, 90, 93, 95, 102, 103, 105, 112, 121, 125, 127, 128, 130, 131, 135, 139, 141, 144, 145, 148, 150, 151, 152, 157, 175, 187
comfort 28, 55, 93, 102, 103, 128, 152, 157
height 44, 121, 131, 135, 152, 175
meter 151
quality 29, 67, 150, 151, 152
Rigid axles 28
Riley 75
Road induced forces 128, 135
input data 144
profile 138
Roll 25, 27, 28, 29, 31, 33, 39, 42, 44, 46, 49, 50, 52, 55, 58, 89, 91, 109, 112, 115, 118, 125, 126, 128, 130, 131, 134, 135, 139, 141, 151, 154, 155, 157, 158
axis 31, 42, 55, 126, 155
centre 28, 31, 50, 55, 58, 89, 91, 112, 158
Rolt, Tony 153
Root mean square 151
Rose joints 118
Rosemeyer, Bernd 112
Rover 65, 73, 74, 76, 102, 131
3500 131
SD1 102
Royal Automobile Club 113
Rubber 35, 36, 44, 50, 93, 95, 97, 101, 118, 128, 130, 131, 157, 159, 169
springs 128, 130
Rumpler, Edmund 109

Run-flat tyre systems 99
rwd 61, 62, 63, 65, 66, 70, 71, 72, 73, 173

Saab 62, 103
Saginaw Steering Gear Division 180
Satellite 86
Scrub radius 30
scs 165
Secondary ride 150
Section moduli 85
Self-aligning torque 29
Self-centring 29, 31
Self-levelling 127, 130, 131, 134, 135, 141, 145, 152
Semi-elliptic leaf springs 19, 105, 109
Semi-trailing arm 50, 51, 52, 53, 157, 158
Semperit tyres 104
Servotronic power steering 184
Shelsley Walsh hillclimb 113
Shock absorbers 27, 38, 148
Silverstone race circuit 113
Simple harmonic motion 37, 38
Sizaire-Naudin car 105
Skirts 125, 134
Skoda 157
Sleeping policemen 42
Sliding pillar suspension 105
Slip angles 29, 90, 155, 158, 159, 163, 183
Snetterton race circuit 113
SP Tyres 99
Space-frame chassis 25
Space-saver tyre 77, 103
Special road surfaces 147
Speedline wheels 101
Split-μ surface 54, 169, 173, 174
Spring rate 26, 28, 38
Springs 11, 19, 25, 26, 27, 28, 29, 31, 34, 35, 36, 37, 38, 40, 41, 44, 45, 49, 50, 51, 58, 105, 108, 109, 112, 115, 121, 125, 128, 129, 130, 131, 134, 135, 138, 141, 144, 152, 154, 175
Stabilia 11
Stability 19, 28, 30, 53, 54, 55, 63, 70, 71, 90, 93, 101, 109, 125, 130, 139, 163, 167, 169, 170, 171, 174
Stabilizer bar 27, 31, 47
Statics 27, 88
Steel 11, 14, 16, 17, 19, 35, 36, 50, 53, 74, 93, 95, 108, 109, 129, 130, 131, 135, 141
Steer 33, 51, 53, 55, 58, 88, 90, 153, 154, 163
Steer-by-wire system 180
Steering control 28, 164, 169, 170, 171, 180
Steering wheel fight 150
Steyr-Daimler-Puch 74, 81
Stiffness 19, 28, 29, 33, 36, 37, 40, 41, 44, 52, 76, 85, 95, 109, 121, 129, 130, 131, 134, 152, 175
Stop control system (scs) 165
Stopping distance 101, 164
Stress analysis 58, 82, 85, 91
Struts 44, 45, 46, 47, 55, 67, 82, 135, 141
Stuttgart 112
Subaru 75
Subjective ride 151
Suspension testing 144
Suzuki Jimmy 74
  SJ410 73
Swing axles 50, 109, 112, 157, 158
Synthetic springs 134

TD tyre system 99
Teves anti-lock brakes 183
Texas Instruments 134
Thermoplastics 97
Thompson, R. W. 93
Toe setting 29, 33, 51
Toe-out 157, 158, 159
Torque boxes 19
Torque steer 31, 33, 89, 90
Torsion bars 27, 31, 35, 36, 47, 130, 181
Toyota Landcruiser 74
Track control arm 47
Tracta cars 61
Trail angle 52, 53, 55, 158
Trailing twist axle 58, 109
Transport and Road Research Laboratory 43
Transverse leaf springs 25
  wishbones 158
Tread life 101
  patterns 95, 97
Triumph Herald 50
Tropfenwagen 109
Trumpf cars 61
TRW Inc. 183
Tubular space-frames 121
TÜV 101
Twin tyres 99, 100, 101, 102
Twin-tube dampers 58, 60, 121, 145
Tyres 25, 28, 29, 30, 31, 33, 34, 50, 53, 54, 55, 70, 71, 88, 89, 90, 93, 95, 97, 99, 101, 102, 103, 104, 118, 128, 138, 151, 154, 155, 157, 158, 159, 163, 164, 170, 171, 181, 183
  footprint 29, 30, 154, 155
  cross-ply 93
  radial-ply 95

Understeer 33, 70, 80, 130, 155, 163
Unequal-length wishbones 45
Uniroyal tyres 104
UNIVAC computers 87
Unsprung masses 28, 37, 38, 51, 55, 108, 127, 139, 141

Vanwall 115
Variable pitch 129
Variable rate springs 129
Vauxhall 17, 53, 164, 165, 171
  Carlton 53, 164
  Cavalier 63
Vectors 28
Vehicle dynamics 28, 140, 154
Vertical forces 41, 127
Visual display unit 82
Voiturette 105
VW Beetle 47, 50, 157
  Golf Synchro 75, 174, 183
  Iltis 74, 75
  Polo 67
  Transporter 77

Wagner, Max 109
Wang computers 87
Watchdog circuit 138
Weight distribution 67
  transfer 130, 139
Weymann bodywork 14
Wheel angle 29, 90
  geometry 25, 51, 126, 128, 152
  offset 30
  tramp 139
Wheel-lock 164, 168, 169, 170
Willys Jeeps 73
Wishbones 45, 46, 47, 49, 50, 51, 55, 109, 112, 118, 121, 125, 129, 158

Yaw angle 28, 51, 89, 90, 151

ZF 180, 184
  Servotronic power steering 183
Zwickau 112

μ 54, 95